SOMERSET LANDSCAPES

Geology and Landforms

by
Simon K. Haslett

Published by Blackbarn Books

First published 2010 by Blackbarn Books, Usk, Monmouthshire, UK.

Website: https://sites.google.com/sites/blackbarnbooks
Email: blackbarnbooks@aol.com

© 2010 Simon K. Haslett

All rights reserved. No part of this book may be reprinted or reproduced or utilised in any form or by any electronic, mechanical, or other means, now known or hereafter invented, including photocopying and recording, or in any information storage or retrieval system, without permission in writing from the publishers.

For Sam, Maya, Elinor and Rhiannon

About the author: Simon Haslett is Professor of Physical Geography and Dean of the School of STEM at the University of Wales. He has published over 100 academic articles and authored/edited several books. He is a Fellow of the Royal Geographical Society and the Geological Society of London, and often appears on television and radio. He is committed to the Public Understanding of Science and regularly gives public lectures to local interest groups.

Follow Simon on:

Blog: http://www.profsimonhaslett.blogspot.com/
Facebook: http://www.facebook.com/pages/Prof-Simon-Haslett/86610699297
Flickr: http://www.flickr.com/photos/profsimonhaslett/
Twitter: http://twitter.com/ProfSHaslett
YouTube: http://www.youtube.com/user/ProfSimonHaslett

Contents

Preface 7

   Chapter 1. Landscapes and Somerset's Geological History 10

      Case Study 1.1 The dynamic earth 11

PART 1 - UPLAND SOMERSET 25

   Chapter 2. The Mendips 25

      2.1 Black Down 25

      Case Study 2.1 Mendips mineral wealth 28

      2.2 Maesbury Castle 31

      Case Study 2.2 Hillslope Processes 34

      2.3 Shute Shelve Hill 38

      Case Study 2.3 Alluvial cones 41

      Case Study 2.4 Tufa - nature's 'limescale' on Mendip 44

   Chapter 3. The Quantocks 48

      Case Study 3.1: Tundra Environments 53

PART 2 - SOMERSET'S HILLS AND VALLEYS 57

   Chapter 4. Somerset's Ancient Hills and Valleys 57

      4.1 Hills of the Axe Valley 57

      4.2 The Yeo Valley 62

4.3 Barrow Gurney 65

Chapter 5. Somerset's Younger Hills and Vales........................... 67

    5.1 Evercreech Vale 68

    5.2 Cadbury Castle.. 72

    5.3 Pennard Hill 76

    5.4 Ham Hill 81

    5.5 Doulting 86

    5.6 Dundry Hill.. 92

    5.7 Somerset's Chalklands.. 97

Chapter 6. Recently Formed Valleys 101

    6.1 Ebbor Gorge.. 101

    6.2 Burrington Combe 105

    6.3 Avon Gorge... 108

    6.4 Blagdon Combe 112

    Case Study 6.1 Dry Valleys 114

    6.5 Chew Valley 116

    Case Study 6.2 Coal Mining at Pensford................................ 118

PART 3 - COASTAL AND LOWLAND SOMERSET 121

Chapter 7. Somerset's Old Coastlines 121

    7.1 The Vale of Porlock... 121

7.2 Brean Down 128

Case Study 7.1 Ammonites of Somerset 133

Chapter 8. Where Lowland Meets the Sea 136

8.1 Saltmarshes and mudflats of the Severn Estuary 136

Case Study 8.1 Tides in the Bristol Channel 141

8.2 Stert Flats 144

8.3 Somerset Levels and sea-level change 148

Case Study 8.2 Detecting sea-level change 153

Case Study 8.3 The 1607 flood – a tsunami in the Bristol Channel? 156

Glossary 161

Bibliography 170

Appendix - Topographical and geological maps of Somerset 184

Preface

This book is intended to be an introduction to the origin and development of various landscapes of Somerset. Most books about the geology of a particular county, region or country usually structure the text so that it documents the geological history of the whole area through time, only making reference to specific locations when required to exemplify the overall picture. This approach has the advantage that the general story of the area is easily followed, but for someone interested in the evolution of the landscape in a particular place it can be quite frustrating, flipping between chapters trying to compile the story for oneself.

In this book, however, I have adopted the opposite approach in that within each chapter particular places are investigated in an attempt to unravel the geological history of those locations, which implicitly contributes to the overall picture. I feel that this approach benefits the lay-reader, who is often mainly interested in the local area, by providing a start to finish account of a particular place. Thus, each chapter is a case study of landscape evolution, examining the rocks, establishing their age, finding out what the environment was like when they were formed (mainly through the use of fossils), and how have they been weathered and eroded to produce the landscape we see today. Of course, the disadvantage in this approach is that the overall picture is not always clear, but to compensate for this, Chapter 1 has been written as a summary to the geological history of the entire county of Somerset and so allows subsequent chapters to be placed in context. The order of the chapters has also been arranged so that they run (very generally) from sites where the rocks are of the greatest antiquity to those of the most recent sediments.

Furthermore, much of the material used in this book was originally written for magazine and newspaper articles, but has in places been re-written, expanded and edited for incorporation into this book. Thus, each section has been written to stand alone, so as to be largely understood without reference to the rest of the book. This, I feel, is greatly beneficial, for in these busy times it allows the reader to dip into the book only occasionally, whenever he or she has the time, rather than read it cover to cover all at once, or to read the sections in almost any order, although it is best to start at the beginning.

This book has been written specifically with the layperson, GCSE, A-Level and undergraduate student in mind; thus, technical terms have to a large extent been avoided. So too has excessive detail, such as the thicknesses of individual rock layers and Latin (specific) names for fossils, although some are given as examples. Where technical terms have had to be used, these are shown in **bold** and defined in the glossary at the end of the book, and *Case Studies* explaining general concepts or phenomena, and how they relate to Somerset, occur throughout. As a consequence, much of the text has been written in a narrative style. The information in this book is based entirely upon the studies of many generations of geologists and geographers (including myself), who have published their findings in academic books and journals. To them I am extremely grateful for all their hard work, without which a book like this would not be possible. Normally when one refers to the findings of these studies in the text, a citation is made (authors name and date of publication); however, this interrupts the flow of the text and, to anyone unused to scientific publications, may make reading difficult. Therefore, I have not included author citation in the main body of the text, nor footnotes, but have compiled a bibliography at the end of the book which includes most of these works.

In writing this book I have been helped by many people, but principally by my family: Sam my wife, and Maya, Elinor and Rhiannon my daughters. They have uncomplainingly accompanied me to most places mentioned in this book, sometimes in cold and/or wet weather, but always with a smile and willingness to help. Also, in the home, time that would normally be spent as a family has been sacrificed, so that I could lock myself away to write. It is only right that this book be dedicated to them. Discussions with my former colleagues in the Department of Geography at Bath Spa University during the 1990s, when most of the text was written, are also acknowledged and thanked. I am indebted also to Roy Smart and Jenny Nicholls of *The Somerset Magazine*, and Jan Brothwell and Andrew Addicott of *Mendip Life* (formerly *Yeo Valley Gazette*) for their encouragement and support over the years, forbearance when I missed editorial deadlines, and for granting permission for me to include previously published material in this book. Finally, thanks to Mike Simms (Ulster Museum) for peer-reviewing the text and making corrections and suggestions to me, but any errors or controversial views that remain are mine alone, and to Sue

Grice for drawing Figures 1.3 and 1.4, and Jonathan Wallen (University of Wales, Newport) for digitising my photographs and making them available on Flickr under a Creative Commons Licence.

Chapter 1. Landscapes and Somerset's Geological History

Landscapes are ubiquitous - we are all part of, or live in, a landscape. Yet there is an incredible diversity of landscapes, with extremes from the frozen tundra of the arctic, to the sweltering deserts of the tropics. Although very different, broadly speaking every landscape is formed in a very similar way. The landscape of Somerset is no exception.

Essentially, landscapes are created through the working of two categories of processes, which come together in a particular combination to produce unique landscapes the world over. Although two landscapes may be similar in appearance, and even formed in similar ways, no two are identical.

Poetic analogies of landscapes are commonplace, such as "the earth is a block of marble shaped at the hands of a sculptor", or "the earth is a canvas upon which an artist has created a masterpiece with the brush". In many ways these analogies aren't too far from the concepts of earth scientists, with the canvas or marble block representing the solid rocks that are evident at the earth's surface, and with the artist's brush or sculptor's chisel representing the power of the elements, such as wind, water, and ice.

One significant difference between these romantic views and those of scientists is that the earth is not considered to be inert like canvas or marble, but a dynamic entity which is full of energy and extremely active. It is from within the earth that the first of the two categories of landscape forming processes originate and because of that they are called **endogenetic processes**.

These processes are driven by energy provided by the earth's internal heat and manifest themselves on the earth's surface in a number of ways. Volcanoes are amongst the most obvious landforms formed by internal processes. Mountains are another, produced by the collision of drifting continents. At first it may not be clear how mountains are a product of endogenetic processes, but continental drift is driven by convecting heat cells within the earth, which sets up conveyor belts that carry the continents. For example, India was once joined to southern Africa and Antarctica, but was split off and drifted north by such a conveyor belt, to collide with Asia, pushing up the Himalayan mountain chain in between.

Case Study 1.1 The dynamic earth

The earth is not an inert piece of rock orbiting the sun, but is a dynamic entity full of activity. The earth is structured in layers with the innermost layer known as the core, which itself is subdivided into a solid iron-rich inner core and a molten outer core. The core is surrounded by the molten mantle, the uppermost layer of which is referred to as the **asthenosphere**. *Finally, the outermost layer is known as the lithosphere, composed of oceanic and continental crust. The boundary between the asthenosphere and lithosphere is known as the* **Moho Discontinuity**, *and is essentially the division between molten and solid rock. The oceanic crust is composed of rocks rich in Silica and Magnesium, and are often referred to as* **SiMa** *(eg* **basalt** *and* **gabbro**)*, whilst continental rocks rich in Silica and Aluminium are known as* **SiAl** *(eg* **andesite**, **granite** *and* **rhyolite**)*. Rocks bearing silica are quite light, and hence metaphorically float to the surface of the earth to form a scum which has solidified to become the crust. At depth, the asthenosphere is silica depleted, and silica-poor rocks such as* **peridotite** *and* **eclogite** *form here. Therefore, a density gradient exists between the asthenosphere, through SiMa, to the least dense SiAl.*

The mantle is hot and fluid and because of this is able to flow. Heat is not distributed evenly through the mantle with some areas hotter than others. At these locations, the hot fluid magma (as it is known) is able to rise which forces cooler magma to descend in order to replace it. This rising and sinking of magma sets up convection currents or cells. The crust bulges where hot magma is rising, eventually leading to the fracturing of the crust which enables magma to escape to form volcanoes. Here new crust is being created and is normally expressed on the earth's surface as a rift, either under the sea like the Mid Atlantic Rift which runs down the length of the Atlantic from Iceland to South Georgia, or on land like the Great Rift Valley in east Africa.

In order to counterbalance the addition of new crust where hot magma rises, crust is destroyed where the cooled magma descends, taking with it mainly oceanic crust down what is termed a **subduction zone***. In this manner, a conveyor belt of crustal movement is set up, and* operates as a series of crustal plates that cover the globe. The continents are carried on these plates, but because they comprise low density rocks, they are seldom subducted. This has given rise to the theory of continental drift and the concept of plate **tectonics**. Each plate is bounded by a combination of constructive, destructive and/or conservative margins. Constructive margins are where stretching of the crust (**extensional stress**) causes crustal divergence through, for example, sea-floor spreading, and new crust is consequently added. Destructive margins are where compression causes crustal convergence. This may result in the collision of continents which may buckle and fold upwards to form mountain chains. Such a phase of mountain building is known as an **orogeny**, and has been responsible for creating almost all major mountain chains, such as the Himalayas. Conservative margins are where plates pass against one another (shear stress), with neither crust being created or destroyed (Figure 1.1).

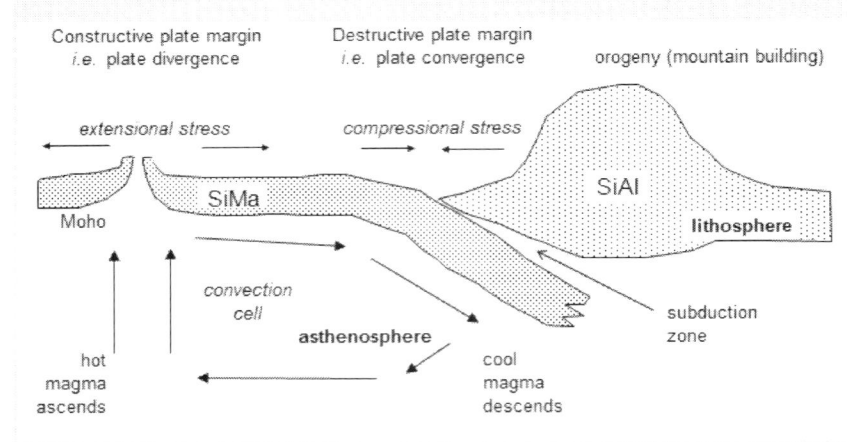

Figure 1.1 A cross-section through the earth's crust showing some of the main features and processes.

The folding of rocks may result in an upfold (convex-up) called an **anticline** or a downfolded (concave-up) **syncline**. If however, the rocks are brittle, they may fracture or fault rather than fold (Figure 1.2). Fault lines are usually sloping within the rocks, with the rocks above the fault plane being termed the **hanging wall**, whilst those below the fault plane are the **footwall**. A normal fault is created where the hanging wall moves down the fault plane relative to the footwall. This occurs under extensional stress and serves to extend the crust. A reverse or thrust *fault occurs where the hanging wall moves up the fault plane in response to **compressional stress** which noticeably shortens the crust. Transform faults are almost synonymous with conservative plate margins, as they describe blocks of rock that move laterally along a vertical fault plane eg San Andreas Fault (California, USA). They can be subdivided into dextral and sinistral types depending on whether the blocks of rock are moving to the right or left respectively.*

Volcanism, continental drift, plate tectonics, mountain building, folding and faulting are all evidence that the earth is a dynamic entity, constantly changing, creating new and destroying old. Furthermore, these processes are crucial in the formation of any landscape, and the Somerset landscape is no exception.

The underlying result of most endogenetic processes, whether in volcanic activity or mountain building, is that the land surface is raised - endogenetic processes generally have a positive (building up) effect on the landscape. However, not all the earth is mountainous. This is because the second category of processes wears the landscape down.

At and above the earth's surface, external processes operate known as **exogenetic processes**. These processes obtain their energy from the sun's heat, and also from the gravitational pull of the sun and moon. They include the effect of moving water (rivers and the sea), ice (glaciers) and wind on the rocks, and they all weather and erode, lowering the landscape. One of the most important solar driven cycles on earth is the hydrological cycle. This describes the transfer of water from different natural stores, such as the ocean, atmosphere, ice caps, glaciers, rivers, lakes, soil water and ground water. The movement of water across the earth's surface, as either running water or ice, is a major agent in landscape evolution, through processes of erosion, sediment transport, and deposition.

Thus, any landscape is the product of the balance struck by the endogenetic processes on the one hand pushing the land up, and the exogenetic processes on the other wearing the land down. In Somerset, this conflict between the two is clearly seen, for we have the Mendips, for example, an area of upland produced long ago by an ancient collision of continents, which are deeply gouged in places by rivers and streams.

Somerset Landscapes

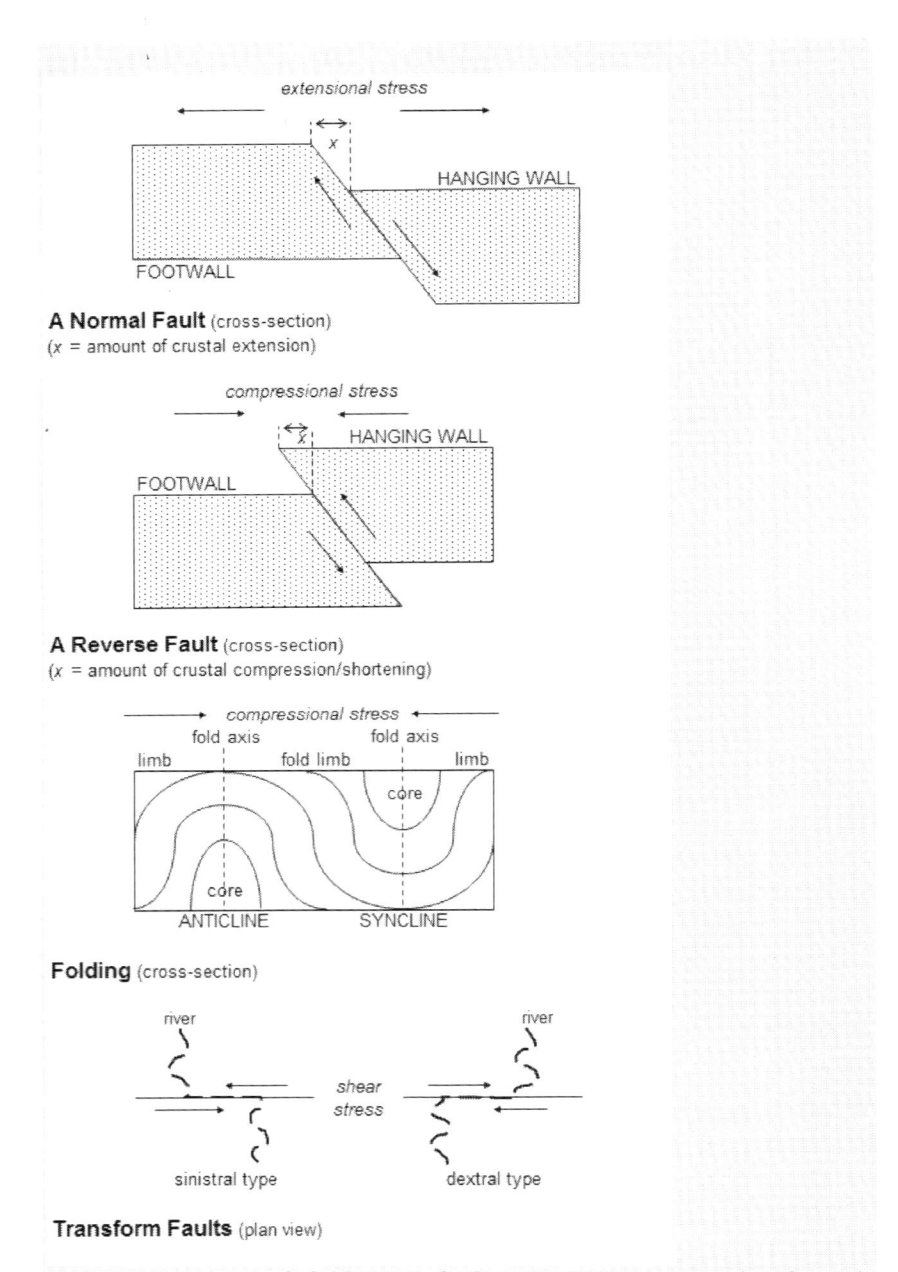

A Normal Fault (cross-section)
(x = amount of crustal extension)

A Reverse Fault (cross-section)
(x = amount of crustal compression/shortening)

Folding (cross-section)

Transform Faults (plan view)

Figure 1.2 Types of folds and faults created as a result of earth movements.

Many British counties are characterised by a certain type of landscape. For example, neighbouring Wiltshire is characterised mainly by treeless flat plains (such as Salisbury Plain). This is because the development of the landscape depends to a great extent on the kind of rocks or geology that underlie a region. The geology of Wiltshire mainly comprises one type of rock in particular, the Chalk. Somerset on the other hand is not characterised by a particular kind of landscape. This is because the geology of Somerset is extremely varied, and each particular rock type, by its nature and degree of resistance it has to erosion, dictates the type of landscape formed. It is this variation in geology that gives Somerset its rich diversity of landscapes, making it the interesting and scenic county that it is (Figure 1.3).

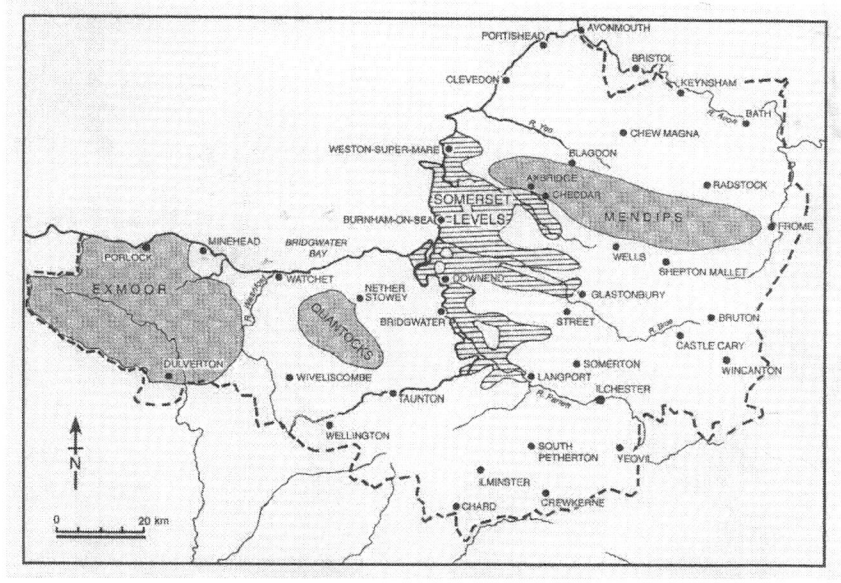

Figure 1.3 The varied topography of Somerset, showing the location of upland (Exmoor, Mendip, Quantocks) and lowland (Somerset Levels).

Most of the rocks in Somerset are known as sedimentary rocks, which are rocks that were laid down or deposited as soft sediment on top of one another. As the layers of sediment accumulated, pressure from the overlying sediments compacted and hardened the underlying deposits to form the hard rocks that we now see. Because sediments are deposited sequentially, the first sediment to be deposited is the oldest and the last to be deposited is the youngest. This principle makes it possible to date sedimentary rocks in relation to one another. Scientists have used this principle to establish a relative time frame against which all rocks can be dated. This time frame is known as the geological column and comprises a number of sub-divisions known as geological periods (Table 1; Figure 1.4).

The rocks that were deposited during each of these periods record the environment that prevailed at that time in what we now call Somerset. The oldest rocks that occur in Somerset belong to the Palaeozoic era (meaning ancient life). During the Palaeozoic, the days were shorter because the earth spun faster, and consequently each year contained more days than at present. The Palaeozoic witnessed an explosion in the diversity of life, including the first plants and animals to colonise land.

The oldest Palaeozoic rocks found in Somerset are of Silurian age (named after a Welsh tribe). These rocks are rare in Somerset and can only be found exposed in quarries on Beacon Hill, northeast of Shepton Mallet. These rocks include **mudstones** rich in fossils, especially seashells and **trilobites** (a primitive animal that crawled along the bottom of the sea) (Plate 1.1). From these fossil animals scientists have been able to suggest that a warm subtropical sea covered Somerset. Other Silurian rocks found on Beacon Hill comprise volcanic lava and ash, indicating that the area was the site of active volcanoes - a far cry from today's tranquil scene.

The Silurian rocks are overlain by deposits of the subsequent period, the Devonian. It is so named because rocks of this age are commonly exposed in Devon. In north Somerset, in the middle of the Mendips, Devonian rocks are known as **Old Red Sandstone**. They are predominantly sandstone and are red because they were generally laid down in a sub-aerial environment rather than under the sea. Because the Old Red Sandstone was deposited on land, fossils are very rare indeed, but some fossil fish have been found in what must be river deposits. It has been suggested that the Old Red Sandstone represents the sub-aerial deposits of a very large delta, stretching from South Wales to Somerset.

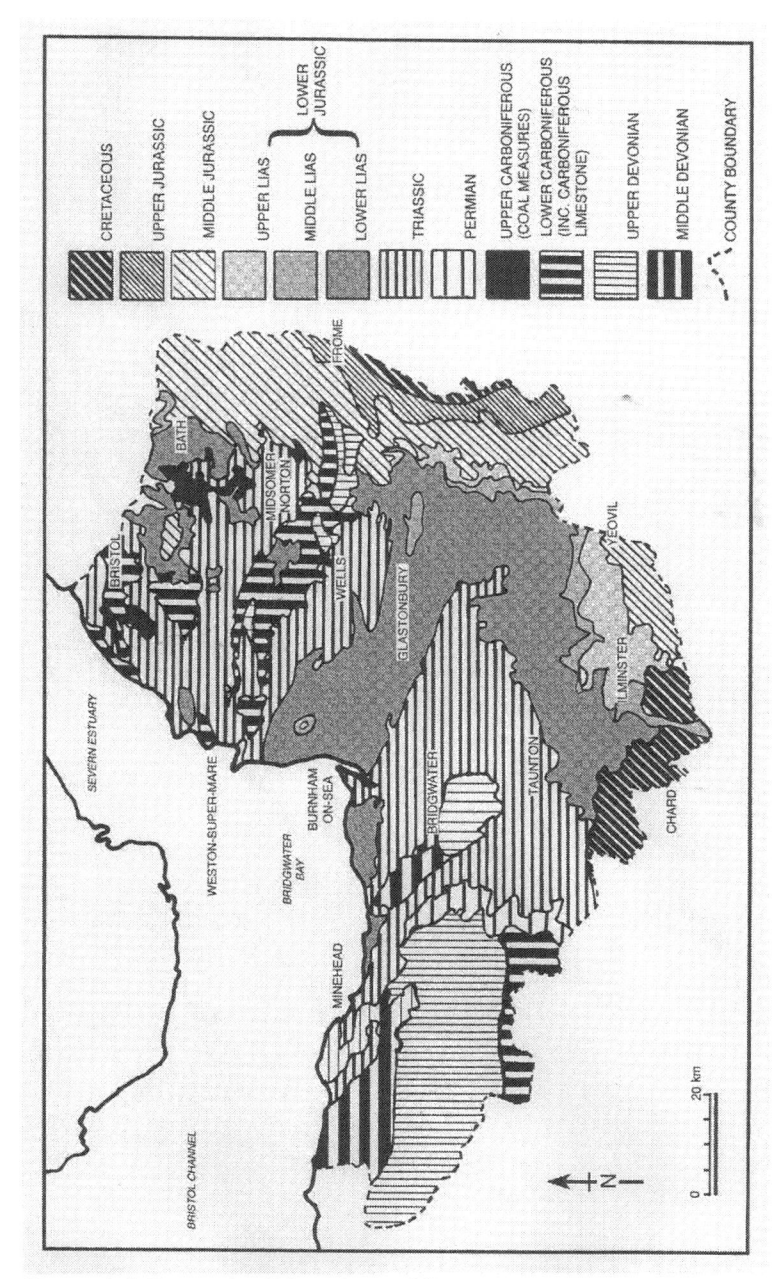

Figure 1.4 General solid geology of Somerset.

Era	Period	Age*	Somerset example
Cenozoic	Quaternary	2	Somerset Levels
	Tertiary	65	Limited occurrence in Somerset
Mesozoic	Cretaceous	145	Chard
	Jurassic	208	Glastonbury
	Triassic	245	Bridgwater
Palaeozoic	Permian	290	Wiveliscombe
	Carboniferous	362	Mendips
	Devonian	400	Quantocks
	Silurian	440	Beacon Hill near Shepton Mallet
	Ordovician	510	None occur in Somerset
	Cambrian	570	None occur in Somerset
	Pre-Cambrian	4600	None occur in Somerset

Table 1.1 The geological column with the age of the beginning of each period in millions of years, and an example of where rocks of each age occur in Somerset (if at all).* The date given refers to the beginning of the geological period and has been obtained through radiometric dating, that is the measurement of the decay (half-life) of radioactive isotopes in the rock.

In west Somerset however, Devonian rocks comprise an array of different types, including limestone, sandstone, slates and volcanic rocks. All these rocks are relatively hard and have formed the upland areas of Exmoor, the Brendon Hills, and the Quantocks. Unlike the Old Red Sandstone of north Somerset, these rocks were deposited in shallow seas and contain fossils of sea-shells and corals. The Devonian shoreline, separating land to the north and sea to the south, is thought to have run east to west through central Somerset and out into what is now the Bristol Channel.

Because of continental drift, Britain at this time was situated very close to the equator and the Carboniferous period, which followed the Devonian, provides us with abundant evidence for Somerset's tropical climate at that time. At the beginning of the Carboniferous the shallow Devonian sea of southwest Somerset started to rise and invade the Old Red Sandstone continent to the north to produce a shallow marine environment very similar to the present day Bahama Banks. The rocks of north Somerset that record this are the **Carboniferous Limestone**'s, rich in fossils of all kinds. These limestones are hard and durable and are quarried extensively throughout the Mendips. Spectacular natural exposures through the Carboniferous Limestone can be seen at Cheddar Gorge.

Plate 1.1 Typical Silurian fossils (includes the large **brachiopod** *Leptaena depressa*, and the trilobites *Dalmanites* and *Encrinurus*).

While the rising sea-level was flooding the Old Red Sandstone continent to the north, the sea that already occupied southwest Somerset simply became deeper. Therefore, the shallow water Devonian deposits of Exmoor can be seen to pass upwards into deep water Carboniferous sediments which contain fossils of deep-sea animals. Most of these deep water rocks occur in Devon, but some can be found in Somerset near the Devon border south of Dulverton.

During the latter part of the Carboniferous the sea-level fell and north Somerset became a swamp forest, and because the wet conditions prevented wood from decaying, these forests became fossilised to form what we now call coal. These coal bearing rocks give the Carboniferous its name (carbon-bearing), and only occur in north Somerset in the Somerset Coalfield around Radstock and Midsomer

Norton, and in the Bristol Coalfield. The Coal Measures, as they are called, yield numerous fossils, including shells and plants, and also fossils of the cockroaches, dragonflies, and "scorpions" that lived in the vast hot humid forests. Writhlington Geological Nature Reserve (near Radstock) is well-known for its fossiliferous Upper Carboniferous rocks, and open days are often organised by the Geologists' Association (Office: Burlington House, Piccadilly, London) where fossil collecting is guided by experts.

Following deposition of the Carboniferous rocks, a period of great earth movement occurred around 290 million years ago, known as the **Variscan Orogeny** (orogeny meaning a period of mountain building). These movements were the result of a southern continent colliding with Britain, uplifting, fracturing and contorting rocks rather like the Indian sub-continent has done more recently to produce the Himalayas on its northward drift. This has folded and faulted the older rocks of Somerset. The Mendips to the north and the Exmoor-Quantocks region to the southwest were uplifted to form new land-masses, exposing the rocks to weathering and agents of erosion. The intervening area was downfolded to produce the low-lying country that is now typified by the Somerset Levels.

Material eroded from the uplifted areas was deposited on the flanks of the uplands during the Permian period. On the slopes of the Mendips and Quantocks for example, rocks occur which contain within them eroded angular fragments of Carboniferous rocks held together by finer red sediment (known as a **breccia**); the angular nature of the fragments suggests that they have not travelled far from where they were eroded, and certainly have not been transported by water, which usually wears down rock fragments to become rounded pebbles (known as a conglomerate when incorporated into a rock). The red colour of the finer material indicates that these rocks were laid down in an arid continental climate.

The area was still above sea-level in the following geological period, the Triassic. The so-called New Red Sandstone was deposited during that time, and blanketed the old eroded Variscan land surface. The environment of these times continued to be hot, and arid desert-like conditions prevailed. Again, like the Old Red Sandstone of the Devonian period, the New Red Sandstone is largely devoid of fossils; however, the remains of early dinosaurs can be found in these deposits. Triassic rocks outcrop over large tracts of Somerset, including the flat lands around Taunton and Bridgwater in the south, and in the area between Clevedon and Congresbury in the north.

Towards the end of the Triassic, sea-levels began to rise, flooding the Triassic deserts with warm shallow seas. These marine deposits comprise alternating layers of limestones and **shales**, which gives rise to the local quarryman's name for these rocks, Lias, and are attributed to the Jurassic period. As many will know from the film *Jurassic Park*, this geological period is famous for its fossils of dinosaurs. Most true dinosaurs however, can only be found in terrestrial deposits (sediments laid down on land), such as the New Red Sandstone, but the marine Jurassic of Somerset has yielded perfect skeletons of the marine reptiles or sea-dragons *Ichthyosaurus* and *Pleisiosaurus*. Indeed, when visiting Street look out for a sea-dragon pictured on the towns' welcome sign. Many other kinds of fossils also occur, including many types of beautiful ammonites for which the Jurassic is famous. These **Lias** rocks occupy a broad area of central Somerset from Burnham-on-sea eastwards to Glastonbury, then south to Marston Magna, and west again to Staple Fitzpaine.

The Lias was deposited during early Jurassic times, whilst in the middle of the Jurassic the shale beds disappear and a continuous marine limestone occurs, known by geologists as the Oolite Series. This is a very fossiliferous rock and yields some the most exquisite ammonites to be found in Somerset. It occurs in part along Somerset's border with Dorset, from Crewkerne to Yeovil and on to Milbourne Port, but from there it forms the hills to the north to Wanstrow and Frome, and then on to Bath. It is from these rocks that the beautiful Bath Stone has been quarried. Overlying the Oolite in eastern Somerset, upper Jurassic rocks occur in a north-south strip of land from Witham Friary through Wincanton and on to Temple Combe. Like the Lias and Oolite these rocks were deposited in a warm sea that covered Somerset at that time.

At the end of the Jurassic there was a minor phase of uplift and erosion. Rocks from the following Cretaceous period were deposited on top of this eroded surface when the sea re-invaded Somerset. Early in the Cretaceous the sea was fairly shallow and two rock-types were deposited, the **Gault** and the **Greensand**. The Greensand is bit of a misnomer, as it is green only when fresh, and turns reddish-brown when exposed to air. Both these deposits contain fossils of sea-shells, including ammonites. Sea-level was rising throughout the Cretaceous, which eventually became deep enough for Chalk to accumulate. Chalk gives the Cretaceous its name (*Creta* from the Greek meaning Chalk), and comprises almost 100% fossils. The fossils which form the Chalk are of marine plankton that lived in the warm sea environment situated on an old continental shelf. Cretaceous rocks are restricted in their occurrence in Somerset, being found only around Winsham, Chard, and the Black Down Hills in south Somerset, and in a thin band along Somerset's border with Wiltshire.

These rocks from the Cretaceous period are the youngest solid rocks found in Somerset and these date from before 65 million years ago. So from that point in the distant geological past to the beginning of the **Quaternary** (ice ages) we have no record of what the environment of Somerset was like. In the intervening period, younger rocks may have been deposited, but have since been largely eroded away.

During the Quaternary ice ages (Table 1.2) Somerset was never completely covered by ice sheets or glaciers, but probably had considerable seasonal snowfields, particularly on the high ground of the Mendips. At any rate, the ground would have been permanently frozen, and an environment very similar to the present day arctic tundra would have persisted until about ten thousand years ago. Products of ice age erosion in Somerset form a deposit known as **head** by the Geological Survey (increasingly known as **colluvium**), and this head can be found in extensive spreads around upland areas. Fossils of mammoths and woolly rhinoceros can be found in these deposits, attesting to their ice age origin. Humans came to live in Somerset during the Quaternary. Some of the earliest evidence of man in Britain comes from a cave at Westbury-sub-Mendip which dates to about 500 thousand years ago. Also, in Gough's Cave at Cheddar Gorge and Avelines's Hole at Burrington Combe scientists have unearthed the remains of reindeer, giant deer, red deer, horse, lynx, ox, pig, brown bear, and wolf in association with human skeletons and flint tools dating from approximately 12,000 and 9000 years ago respectively.

Epoch	Date began (years ago)	Stage	Climate	Evidence in Somerset
Holocene	10,000	Flandrian	present interglacial	peat and alluvium in the Somerset Levels
Pleistocene	122,000	Devensian	last glacial	head deposits and alluvial cones
	130,000	Ipswichian	last interglacial	Burtle Beds and raised beaches
	300,000	Wolstonian Complex	a series of cold and warm stages	uncertain evidence in Somerset, perhaps some river terraces
	340,000	Hoxnian	interglacial	uncertain evidence in Somerset
	475,000	Anglian	major glacial	widespread glaciation of Britain, including parts of Somerset
	502,000	Cromerian	interglacial	uncertain evidence in Somerset
	7550,000	Beestonian	glacial stage	mainly periglacial, limited evidence for glaciers in Britain
	1,810,000	Early Pleistocene stages	a series of cold and warm stages	no evidence in Somerset

Table 1.2 The Quaternary Period, its epochs and stages, with dates for the beginning of each stage, and comments on evidence for each stage in Somerset.

Following the melting of the ice sheets, sea-levels rose quickly and flooded western and central Somerset. Indeed, the whole area which we now call the Somerset Levels is underlain by estuarine clay and peat from this time. The sea occupied this area until Roman times, and if global warming becomes a reality these areas could be flooded again.

The landscape we see in Somerset is a product of millions of years of deposition and erosion, and these processes continue today. Geology may be regarded as the canvas upon which the landscape picture is painted; the characteristics of individual rock-types and their structure determines the manner in which they are carved up by the elements of erosion - water, ice and wind, ultimately controlling the kind of landscape that evolves. Somerset is lucky in that it has a varied geological history which makes an incredible story in itself, and has created the wonderful diversity of landscapes that makes Somerset such an attractive county. Geology occurs under everyone's feet and is free to be explored by everyone. Regardless of your age or where you live, geology is inexpensive to study, and above all, if you observe health and safety guidance, it is fun and stimulating.

PART 1 - UPLAND SOMERSET

Chapter 2. The Mendips

The Mendips are an upland area in north Somerset, stretching in a line from Frome in the east to Weston-super-Mare in the west. They rise very abruptly from the surrounding countryside to heights greater than 300m above sea level. The rocks of the Mendips are very old and have been contorted and folded by earth movements millions of years ago. The geology of the Mendips is illustrated here by reference to three localities, two from the central part of the upland (Black Down and Maesbury Castle) and the other from a position on the flanks of the upland (Shute Shelve Hill) (see Figure 2.1).

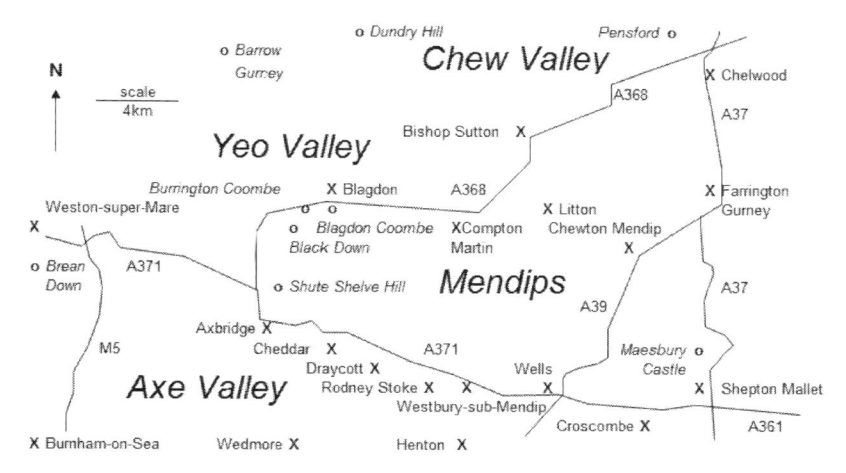

Figure 2.1 Location of some of the places mentioned in the text from the Mendips, Axe, Chew and Yeo Valley areas.

2.1 Black Down

Black Down rises to some 325m at Beacon Batch in the Mendips. It is a high upland area of bracken and isolated trees which towers above the village of Burrington (Plate 2.1). Yet beneath the rolling hills of Black Down lie rocks contorted by the forces of an ancient collision of continents.

Somerset Landscapes

Plate 2.1 A view to the north from the summit of Black Down, overlooking Burrington Combe, and the Yeo and Chew Valley's beyond.

At the summit of Black Down, fragments of a red coloured rock known as Old Red Sandstone can be found. The Old Red Sandstone was deposited in a deltaic environment during the Devonian geological period between 400 and 362 million years ago. But down the slopes of Black Down and into Burrington Combe, the rock type changes to a blue-grey coloured limestone which was deposited in warm tropical seas of the Carboniferous period, some 350 million years ago. This Carboniferous Limestone is well exposed in the valley sides at Burrington Combe, where fossils of corals and sea-shells can be found.

From the centre of Black Down to its northern edge, the rocks become progressively younger. This means that these rocks are no longer lying flat as they were when first deposited, and are now inclined towards the north. In Burrington Combe, rock surfaces can clearly be seen to incline or dip steeply towards the north. On the southern side of Black Down, the rocks there too are dipping, but this time to the south, so they still become progressively younger away from the middle of Black Down.

The summit of Black Down forms the middle or core of a geological structure called an anticline (Figure 2.2), whereby the rocks have been deformed or folded, from their original horizontal position, into a dome-like structure, which in this case is rather asymmetric. The inclined rocks on either side of the core of an anticline are called its limbs. Anticlines are formed by compression, for example, by two continents colliding into one another, so compressing and buckling the rocks that lie in between. It is worth mentioning, that because the fold axis of the Mendip anticline plunges into the ground, it is more accurately termed a **pericline**.

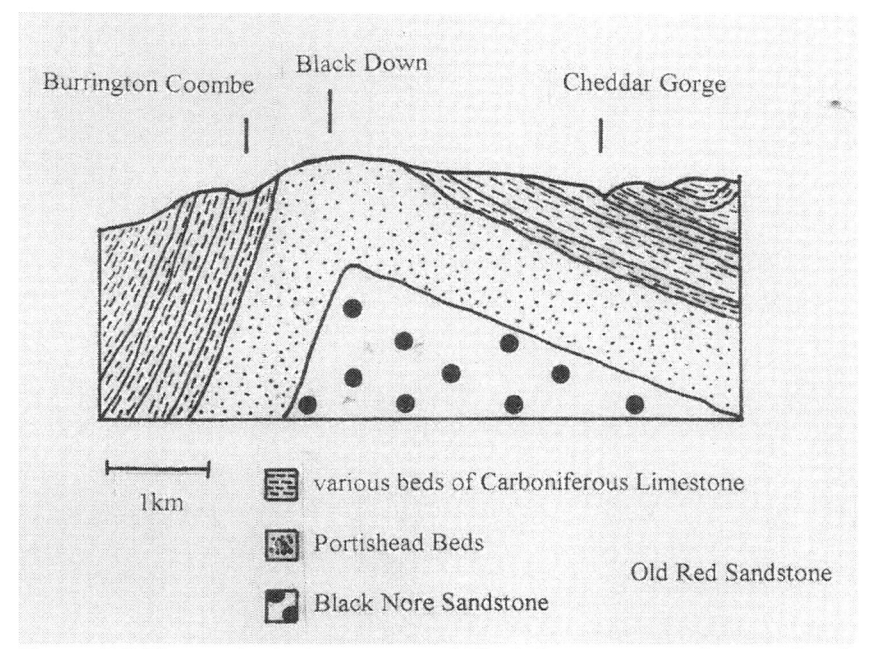

Figure 2.2 The geological structure of Black Down - a good example of an anticline.

The Black Down anticline is unusual, as most anticlines, ironically, form valleys. This is because the core of an anticline usually fractures under compression and is weakened, and so is more prone to erosion than the limbs. By analogy, if one bends both ends of a Mars Bar downwards, for example, the chocolate in the middle (core) of the bar would crack and probably fall off, whilst the chocolate on the ends (limbs) of the bar would remain intact. The reason for Black Down's resistance to erosion is probably because in the present climate the Old

Red Sandstone resists erosion more effectively than the Carboniferous Limestone, which is susceptible to being dissolved by water; thus, water flows off the Old Red Sandstone and onto the Carboniferous Limestone, which is consequently eroded faster.

The event that compressed and folded these rocks is called the Variscan Orogeny, and took place at the end of the Carboniferous around 290 million years ago. It occurred when a continent to the south collided with a northwest European continent. The actual zone of collision, where the two continents met, is situated across what is now central France, so that the Mendips are a long way to the north. Yet despite the distance the force of the collision was such that it was strong enough to deform and fold rocks many miles away in a place now known as Black Down.

Case Study 2.1 Mendips mineral wealth

Mendip is not only justly famous geologically for its limestone scenery of pavements and gorges, but also for the mineral wealth that has been exploited locally since before Roman times. The reason why Mendip is so endowed with commercially valuable minerals lies in the geological history of the area, and its geological structure and the rock types that make up the Mendip Hills.

*The majority of rocks that are found on the Mendips are sedimentary rocks, that is they were deposited originally as soft and unconsolidated sediment, such as gravel, sand and mud. With continued deposition these sediments became squashed and pressurised by overlying sediment, and underwent **lithification** and **diagenesis**, the processes that convert soft sediment into hard rock. Mineral particles may have been deposited along with all the other sediment and, in this case, the minerals would be distributed randomly throughout the rock. Such mineral concentrations are called placer deposits.*

Plate 2.2 The mining landscape near Priddy has been exploited for its mineral wealth at least since Roman times.

On the Mendips however, the main occurrences of minerals, found in central parts of the Mendips (for example, Priddy, Plate 2.2) are highly concentrated and so very economical to extract. The main minerals for which the Mendips are famous are lead and zinc ores, **galena** and **sphalerite** respectively. Because these ores are so highly concentrated they are unlikely to have been deposited with the sedimentary rocks, and were probably emplaced at a later time.

Such ores are commonly transported dissolved in hot water, which has risen from deep down in the earth's crust, and then precipitated out as temperature and pressure decreases as the water nears the earth's surface. This emplacement of minerals is called **hydrothermal mineralisation** because of the involvement of hot water. Not unsurprisingly this water travels through joints and fractures in the Carboniferous Limestone, and through pore spaces between sand grains in the Devonian Old Red Sandstone. It is the precipitation of ores in joints and fractures that forms highly concentrated mineral veins. But why are they so abundant in the central Mendips?

After the sedimentary rocks were deposited, mainly during the Devonian and Carboniferous geological periods, the area was affected *by earth movements related to continental drift which folded Mendips rocks into an upfold called an anticline. Therefore, all around the Mendips the rocks slope (or dip) up to the centre of the Mendips, the core of the anticline. As the hot waters bearing the dissolved minerals rose through the earth's crust they followed the up-sloping layers of permeable Carboniferous Limestone and porous Old Red Sandstone, and were essentially channelled from all directions towards the centre of the Mendip anticline where the minerals were ultimately precipitated to form highly concentrated veins of ore.*

*From studying the occurrence of different hydrothermal minerals in various rocks on Mendip, we know that there were at least two episodes of mineralisation. The first episode occurred following the Carboniferous period, during the Triassic and Jurassic periods. It was this episode that emplaced most of the economically important lead, zinc and iron ores. The second episode occurred after the Jurassic period, but these hot waters notably contained silica, which has **silicified** a number of Triassic and Jurassic rocks, which are widespread on the top of the Mendips, and have become known as the **Harptree Beds** (Plate 2.3).*

Plate 2.3 High Mendip plateau (approximately 220m above sea-level) at Shooter's Bottom Farm (near Emborough). The ground here is partly underlain by the silicified Harptree Beds.

It is quite amazing to consider that these mineral ores represent the movement of water from the interior of the earth to the surface, bringing with it in solution materials common at those depths, to produce minerals that very valuable here on the surface. A testament to the dynamic nature of our earth.

2.2 Maesbury Castle

Maesbury Castle is an impressive Mendip hillfort situated near Shepton Mallet. Iron age people undoubtedly chose this hill as the site for their settlement because of its elevation, drainage and topography. All these attributes are due to the geology of the area; the rock type, geological structure, and the way in which these two factors interact with weathering and erosion to produce the landscape we see today.

Maesbury Castle rises to 292m, nearly 1000ft, above sea level. It rises head and shoulders above the rolling hills around Shepton Mallet to the south, which are approximately 100m lower. To the north, the Mendip Plateau is more in the region of 50-70m lower; for example, Oakhill to the northeast is some 248m above sea level, whilst Green Ore to the northwest is around 227m high. Therefore, good vistas throughout almost 360° around Maesbury Castle are the norm (except when the cloud is low!) (Plate 2.4). Although a public footpath traverses through the middle of Maesbury Castle, access is not easy. This is mainly because of the lack of parking on the narrow roads near the Castle.

The reason why Maesbury Castle hill is upstanding and topographically prominent is because of the rock type that underlies the immediate area. Most of the Mendip Hills are composed of Carboniferous Limestone. As with all limestone, the Carb Lime (as it is affectionately referred to by geologists) dissolves easily in weak acidic rainfall. This weathers the limestone away relatively rapidly, producing many of the well-known Mendip geological and landscape features, such as caves, **swallow-holes** (**swallets**), and gorges. Therefore, to produce an upstanding area of rock, such as Maesbury Castle hill, requires a different rock type.

Plate 2.4 A view from Maesbury Castle of the lower surrounding landscape.

The rock underlying Maesbury Castle hill is not a limestone, and so does not dissolve. It is chemically inert. In fact, it is a sandstone belonging to a geological series of rocks collectively referred to as the Old Red Sandstone. As already mentioned, the Old Red Sandstone was deposited in the geological period called the Devonian some 400 million years ago. Being older than the Carboniferous Limestone, it underlies it (that is, the Carboniferous Limestone was deposited on top of the Old Red Sandstone). Although this sandstone does not weather chemically, it does undergo erosion through streams flowing across its surface, cutting down into it and taking eroded material away. However, this process is slower than the chemical weathering of the Carboniferous Limestone that encircles the hill, leaving it upstanding.

The precise name given to the Old Red Sandstones found beneath Maesbury Castle is the **Portishead Beds**. These are so named because they were first described around Portishead near Bristol, where they are exposed in cliffs. They comprise both sandstone and a coarser component known as a conglomerate. A conglomerate is any rock which contains rounded pebbles. In this case, the pebbles are made of quartz, and it is quite common to find these quartz pebbles lying around on the ground, as they persist long after their sandy matrix has been lost. The Portishead Beds constitute the Upper Old Red Sandstone in the region.

The origin of these deposits has been the cause of some controversy. In the past, geologists have suggested that the sandstones represent fossilised sand dune deposits laid down in an arid desert environment. However, more recently, a consensus has developed suggesting they represent deposition in river systems, perhaps in a delta setting. This would explain the occurrence of conglomerates, the constituent rounded quartz pebbles being river worn, and the presence of some fossils. In the Avon Gorge where Portishead Beds are well-exposed, two layers of fossil plants have been found, which it is suggested grew by the banks of rivers. The reason for the considerable geographical extent of these deposits lies in the probability that very great rivers indeed were involved. Some authors believe the rivers responsible for the deposition of the Old Red Sandstone had their headwaters in northeast Scotland!

I indicated earlier that the Portishead Beds occupied the central part of Maesbury Castle hill, with the flanks underlain by Carboniferous Limestone. This is possible because both the Old Red Sandstone and Carboniferous Limestone beds have been folded. Folding occurs when pressure is applied (you can demonstrate this by holding apart both ends of a sheet of paper, and then bring your hands together - the paper folds just as rocks do). The pressure that is needed here to create these large folds is produced by two continents colliding into one another. In this case an ancient southern continent, of which the south of France was part, collided with the continent carrying Britain. All this happened some 290 million years ago, and this collision created the pressure that buckled the rocks. This particular collision event is of course the Variscan Orogeny. Two major types of fold are created in this way, an upfold called an anticline, and a downfold known as a syncline. In the case of Maesbury Castle hill, we are dealing with an anticline, with the older Old Red Sandstone at its core, and the fold limbs on either side comprising younger Carboniferous Limestone.

Unfortunately there aren't many exposures of the Portishead Beds in the vicinity, but plentiful rock fragments can be observed and a few collected from the ground where they have come to the surface. Some of these fragments have been brought to the surface by the digging of earthworks by Iron Age settlers on the hill. Both ditches and ramparts occur circling the summit of the hill. Apparently, the settlement developed relatively late in the Iron Age period and also appears to have been paired with another hillfort on Blackers Hill to the north between Gurney Slade and Chilcompton. In more modern times, sheep tracks have also churned up the ground revealing new rock fragments.

The hill on which Maesbury Castle stands is a wonderful feature in the landscape, rising fairly abruptly from Shepton Mallet in the south, to a summit that overlooks the countryside for many miles in nearly all directions. It is unusual in the Mendips, in that it is formed from Old Red Sandstone and not the otherwise ubiquitous Carboniferous Limestone. Its existence is solely attributable to the properties of this rock type and the anticlinal structure the rocks have assumed. It is a persuasive testament to the influence of geology in the development and evolution of the Somerset landscape.

Case Study 2.2 Hillslope Processes

Like all upland areas, Mendip is defined by its slopes. In some places these slopes may be near- vertical or even cliffed and characterised by bare rock, but in many cases the slopes are less dramatic and are covered by soil and, where the soil is deep enough, small plants and trees. All these are slopes and are mobile, to varying degrees, because of the effect of gravity. These slope movement processes can have an impact on people living on or working the slopes and on man-made structures built near slopes, such as roads.

There are three main types of slope movement active in present day Mendip: landslide, soil heave, and rock fall. During the last ice age other types of movement such as mudflows may have occurred when the soil was waterlogged from seasonal melting of ice. The type of movement that occurs depends largely on the steepness of the slope because on slopes with angle greater than 25° the soil cover is thin or absent.

Rock falls are characteristic of near-vertical slopes and cliffs, such as can be seen in Burrington Combe and Cheddar Gorge. Falls occur when rock becomes detached from the slope face by weathering (including the action of rock climbers) and simply falls to the foot of the slope under gravity. If falling rocks impact on other weathered rock below, then further rock falls may be triggered. Another type of rock fall is called toppling, whereby slabs of rock topple away from the slope face, but with the base of the slab remaining in position. This is not very common on Mendip.

Landslides (Plate 2.5) are fairly common along Mendip slopes and examples may be seen south of the A368 between Compton Martin and Blagdon, and on the slopes of Red Hill near Chelwood. However, these examples are not like the landslides that people usually perceive, but are relatively small and generally do very little damage, although a disruptive landslide in 1995 obstructed the A37 just south of Pensford for some time. The main type of landslide to occur in the area belongs to a class of landslides known as rotational slides (Plate 2.6), whereby a block of soil slides over a bowl-shaped plane of weakness under the soil to produce a depression in the hillside, often marked by a small cliff or scarp of exposed soil at the back of the slide. The material that moves downslope forms a lobe of soil and many cracks may appear on the surface between the lobe and scarp, indicating tension in the soil.

Soil heave is by far the least obvious type of slope movement, but unlike rock falls and landslides which occur episodically, heave is operating almost continuously. It involves the downslope movement of soil through the expansion and contraction of the soil layer. When soil gets wet it expands and pushes the ground surface out at right angles to the slope. When the soil then dries out it contracts, however the ground surface does not return to its original position, as the effect of gravity forces it to contract vertically downward, so moving the ground surface and soil downslope. Essentially the soil heaves up and down and moves downslope. Soil heave is most active where the soil is rich in clay (as is the case in Mendip) because some clays expand greatly when wet.

These different types of slope movement may occur in isolation, but often they occur together to produce distinctive features. For example, **terracettes**, *which are stairlike features commonly seen on fairly steep slopes, are thought to reflect a combination of soil heave and small-scale rotational sliding (Plate 2.7).*

Somerset Landscapes

Plate 2.5 A small landslide (debris flow) on the steep slopes above Porlock.

Plate 2.6 Rotational slides on the flanks of Wearyall Hill, near Glastonbury.

Plate 2.7 Terracettes on the slopes of Cley Hill, near Frome.

2.3 Shute Shelve Hill

Shute Shelve Hill lies towards the western end of the Mendip Hills; it rises to some 233m high, and forms part of an important geological structure in this area. It is an excellent location for studying geology, and collecting rocks and fossils, because it is managed by the National Trust, with good access and many small exposures of the country rock.

The high summit of Shute Shelve Hill is a prominent landmark on the Mendip skyline. Its slopes plunge dramatically to the west and the A38, and to the south, with Axbridge situated on the footslopes, whilst to the north the slope is a little more gentle toward Hale Combe and Sidcot. To the east, Shute Shelve Hill merges with other Mendip Hills, Fry's Hill and then Callow Hill. The Hill is most easily accessed from the A371, where a car park and information boards have been provided. Footpaths are well sign-posted, but a stout pair of shoes is needed to cover the often rough terrain.

The views from Shute Shelve Hill are numerous and include a magnificent vista up the Axe Valley to Cheddar and Nyland Hill, across the A38 to the bracken topped ridge of Cross Plain (Plate 2.8), and northwest to the bleak heights of Wavering Down and beyond to the Lox Yeo River Valley. On a fine day the climb up Shute Shelve Hill is rewarded by these scenes alone. Yet to the geologically interested layperson, Shute Shelve Hill has much more to offer.

The rocks underlying Shute Shelve Hill are sedimentary rocks deposited during the beginning of the Carboniferous Period and are assigned to the **Dinantian Series** of rock some 350 million years old. The rocks themselves are all various types of limestone collectively known in this part of Britain as Carboniferous Limestone, and there are four distinct layers of this limestone exposed on Shute Shelve Hill.

The oldest layer is known as the Lower Limestone Shale and is exposed on the northern flank of Shute Shelve Hill where it overlies the older Old Red Sandstone of the Devonian geological period near Hale Combe. The Lower Limestone Shale comprises alternating shales and limestones rich in fossils of sea animals, such as **crinoids** (sea-lilies) and brachiopods (a type of sea-shell). And so the junction between these two rock-types represents a major change in environment from continental to marine. Occasionally in the Mendips, a thin layer of sandy limestones, known as the 'passage' beds, marks the transition between these two environments, which also contain fossils, including sharks teeth.

Plate 2.8 A view of Cross Plain from Shute Shelve Hill.

Further up the north flank of Shute Shelve Hill, the Lower Limestone Shale is overlain by another layer of limestone called the Black Rock Limestone. This layer comprises dark grey to black hard limestones, sometimes with thin bands of **chert** (a rock made of silica very similar to flint). Fossils of corals and brachiopods are often abundant, and in some of them their original shells have been replaced by silica which leaves them beautifully preserved. These fossils suggest this limestone was deposited in a warm-water coral reef environment.

At the summit of Shute Shelve Hill, the Black Rock Limestone is overlain by a fairly uncommon type of limestone. Here it is called the Burrington Oolite and can be seen exposed in small disused quarries near the summit (Plate 2.9). Oolitic limestones are made up of very small spheres of lime known as ooids. Ooids are forming today in the Bahamas where currents wash the small particles of lime back and forth on the seabed, precipitating lime concentrically around the particle in the warm water, so producing a little sphere or ooid. It is thought that the Burrington Oolite was deposited under similar conditions. However, the movement of the sediment that was required to create ooids also made life difficult for would-be fossils, hence the Burrington Oolite is not particularly rich in fossils. Furthermore, any fossils that are recovered (mainly corals and brachiopods) are often seen to be abraded, probably by these same water currents.

Plate 2.9 Strewn blocks of Burrington Oolite on the summit of Shute Shelve Hill.

On the lower slopes of the southern flank of Shute Shelve Hill, the youngest variation of Carboniferous Limestone outcrops. It is called the Clifton Down Limestone and can be seen exposed in a small overgrown quarry close to the car park. It comprises dark grey to black muddy limestones with occasional beds of **oolite**. Again, it is fossiliferous and fossil corals and brachiopods may be found here.

From north to south, up and down Shute Shelve Hill, it is possible to traverse a small span of geological time from the oldest beds (Lower Limestone Shale) to the youngest beds (Clifton Down Limestone) of the exposed Carboniferous Limestone (Table 2.1). The reason why this is possible lies in the fact that these layers of limestone, which were originally deposited horizontally, are now tilted towards the south. However, on the northern side of the Mendips, such as at Burrington Combe, these same limestones are tilted northward. This is further evidence for a major fold in the rocks of this area, which takes the form of a dome-like geological structure known as an anticline. Here the top of the dome has been eroded off to expose increasingly older rocks towards the centre, or core, of the anticline. Either side of the core the limbs of the anticline occur, and Shute Shelve Hill occurs on the southern limb of this anticline.

Series	Epoch	Formations	Fossil zonation
Carboniferous Limestone	Visean	Hotwells Limestone	*Dibunophyllum* (D) zone
		Clifton Down Limestone	*Seminula* (S_2) zone
		Burrington Oolite	*Caninia* (C_2) & *Seminula* (S_1) zones
	Tournasian	Black Rock Limestone	*Zaphrentis* (Z) & *Caninia* (C_1) zones
		Lower Limestone Shale	*Vaughania* (K, syn. *Cleistopora*) zone
Old Red Sandstone		Portishead Beds	

Table 2.1 A summary of the geological succession of the Mendips.

Again, this folding is the result of the major episode of mountain building, the Variscan Orogeny, which is responsible for creating most of the upland areas of southwest Britain, northwest France, and northeast North America. It occurred before the Atlantic Ocean was created, when Europe and North America were still attached to one another. At this time, a piece of continental crust drifted northward to collide with another piece of crust to push up mountains. It was this collision that compressed the rocks that we now see on Shute Shelve Hill, causing them to buckle and fold to become part of an important geological structure, the large anticline of the western Mendips.

Case Study 2.3 Alluvial cones

The surface of the Mendips today is largely devoid of rivers and streams. This is because the Mendips are widely underlain by rocks that allow water to percolate downward into the earth rather than flow over the surface. Yet in some places on the Mendips there is evidence that at some time in the past torrential water-courses flowed over the Mendip Hills.

The commonest rock type found in the Mendips is the Carboniferous Limestone. Limestone is composed of calcium carbonate or lime, which is very easily dissolved. For most of us living in the Mendips hard water is a problem, especially furring up kettles, necessitating regular descaling. This somewhat annoying disadvantage of living in the Mendips is due solely to geology and the presence of limestone through which our drinking water travels, dissolving the limestone as it goes. Thus, initially narrow joints and bedding surfaces within the limestone are rapidly enlarged by dissolution to allow nearly all surface water to percolate downwards, rendering the limestone highly permeable. In this way, the world famous caves of the Mendips have been formed.

However, at a number of locations around the Mendips there exist substantial bodies of geologically recent sediment that indicate that rivers, and of some magnitude, must have flowed across the Mendips. These bodies of river sediment comprise boulders, pebbles, sand, silt and clay, with the largest components indicating that the rivers must have been fast flowing with a considerable volume of water. Also, this river sediment occurs in discrete cone shaped landforms which have become known as alluvial cones or fans (alluvium is any sediment deposited by moving water).

The occurrence of these alluvial cones appear somewhat enigmatic, but may be explained when we consider the environmental history of the Mendips over the last two million years. During this time the earth experienced several ice ages, all of which affected the Mendips to varying degrees. On at least one occasion an ice sheet lapped onto the north and west flanks of the Mendips, but there is no evidence for glaciers on the Mendips themselves. Instead, the Mendip environment would be that of tundra-like conditions, with dwarf vegetation, and permanently frozen soil, called permafrost. It is during these cold episodes, with the joints in the limestone sealed by permafrost, that water could flow over the Mendip surface. Also, spring snowmelts would feed swollen rivers, cascading down the Mendips, and able to transport large boulders.

This alluvial sediment accumulated as a cone where a steeply inclined river met more gently sloping land, so that the velocity of the river decreased as the gradient decreased. With reduced flow the ice age rivers were no longer able to move larger boulders and pebbles and they were dumped. In the Mendips, alluvial cones occur in such situations, for example, on the Somerset Levels at the mouth of Cheddar Gorge, where it is now built upon, around Litton, where the River Chew descends steeply from Chewton Mendip to the lower area now occupied by Litton Reservoir, and underlying Wells. It is quite fascinating to think that the tranquillity of the upper Chew Valley, the silence of Cheddar Gorge, and the bustling of Wells city centre were at one time replaced by torrential ice age rivers, swollen by snowmelt, and laden with boulders (Plate 2.10).

Plate 2.10 The River Sheppey at Croscombe, the benign descendant of a torrential ice age river.

Case Study 2.4 Tufa - nature's 'limescale' on Mendip.

The majority of the Mendip landscape is underlain by limestone, mainly hard Carboniferous Limestone, so named because it was formed in the geological period called the Carboniferous, which ended some 290 million years ago. And the area surrounding Mendip is underlain by younger Jurassic limestones. A lot of limestone!

The significance of this is that limestone is chemically made up of calcium carbonate. This is an alkali and easily dissolved by rain water, which is naturally slightly acid. This acidity is nothing to do with the 'acid rain' of recent environmental concern, but is due to the natural uptake of carbon dioxide (CO_2) by rain in the atmosphere to form a weak carbonic acid. It is this natural acid rain that has, through geological time, dissolved Mendip limestone, producing 'hard' water, and so contributing to the formation of swallets, caves, caverns and gorges, that Mendip is so justly famous for.

Often the dissolved limestone is carried in the underground waterways below Mendip, but eventually re-emerges via springs. If upon resurfacing the water loses its CO_2 back into the atmosphere, then the water will become less acidic and so redeposit the limestone by precipitating the calcium carbonate. This new 'limestone' is called **tufa** or travertine, and is usually much softer than the original limestone. Such deposits are marked as calcareous tufa on the geological maps that cover the region, but extensive examples can be seen near Rodney Stoke and Ston Easton, and in streams around Coleford (Plate 2.11).

Tufa can assume many physical forms depending on the nature of the spring or the stream fed by the spring. Those on steep slopes often form 'waterfalls' of tufa, known as cascades, whilst water collecting on flat surfaces can give rise to 'ponded' or paludal tufa. Quite often tufa is white and quite crumbly to handle; therefore, easily identified.

The loss of CO_2 direct from the water can occur through processes such as outgassing or degassing (both terms are fairly self-explanatory). This can happen if the water warms up, which is commonly the case when a subterranean stream resurfaces. This is analogous to boiling a kettle of 'hard' water which leaves limescale on the filament. However, this mechanism of tufa formation is often very slow, but is accelerated *where water is vigorously mixed, such as in rapids or a waterfall, and so may explain the formation of tufa cascades.*

Plate 2.11 General view of a tufa cascade in a stream at Coleford.

It is well known that plants, during photosynthesis, require CO_2. Plants that may be partly or fully submerged do take CO_2 directly from *the spring-fed water, and fairly rapidly. The more the plants grow, the more CO_2 they extract, the lower the acidity becomes, and the potential for tufa to deposit increases. Therefore, it is not the chemical composition of the water that determines tufa deposition, but the scale of carbon dioxide removal. In this way, of two springs of similar water chemistry, only one may have associated tufa.*

Importantly, especially regarding paludal tufas, this photosynthesising flora includes aquatic algae, such as diatoms. The algal cells, and other silt particles, actually have tufa deposited around them, so when they fall to the bottom of the stream, pond or marsh, their accumulation becomes tufa! Also, animals grazing on these algae will accelerate this process by producing lime-rich droppings or faecal pellets. Because algae are seasonal bloomers, usually in the Spring-time, it is known that some tufa forms seasonally too, indeed, such is the rate of tufa production at these times, that the water can turn milky white, events which are referred to as whitings! Tufa can accumulate in this way over thousands of years and form extensive deposits. These often are, or can become, vegetated and often support rare flora and fauna, including orchids. Active tufa sites have been singled out for special attention by conservationists and in Somerset a special report has been published by the Somerset Environmental Records Centre of the Somerset Wildlife Trust which outlines the importance of Somerset's tufa sites for local biodiversity.

Like many sediments, tufa can preserve fossils. This is especially true of snails, whose shells are also made of calcium carbonate. By studying tufa's and their fossils it is possible to piece together environmental histories for local areas around tufa sites. For Mendip this is very important, because there are very few other sources for this information. At Bath Spa University a number of colleagues and myself initiated the Mendip Landscape Project, whose focus was to unravel the long-term history of the Mendip landscape, mainly over the last 10,000 years, and principally through studying tufa deposits. This period spans the Palaeolithic (ancient stone age) to the present day, and already there appears to be an enduring relationship between people and tufa sites throughout prehistory, including a stone tool found at Ston Easton. Perhaps the springs were of practical use to prehistoric people for water supply, but maybe they were also of ritual significance. For example, it is not difficult to appreciate the symbolism between birth and the emergence of a spring from the Earth, especially if the water turned to milk every Spring-time!

Chapter 3. The Quantocks

The Quantock Hills are an upland area situated in west Somerset, separating the Somerset Levels to the east from Exmoor and the Brendon Hills in the west (Figure 3.1). They run for approximately 13 miles, aligned northwest-southeast. Although quite rounded on top they are, nevertheless, rugged and bleak and afford splendid views of much of Somerset, the Bristol Channel, and Wales (Plate 3.1). The Quantocks and the area surrounding them are considered to be a classic geological area. It has long been studied by geologists, including some of the pioneers of British (and world) geology, Henry De La Beche and Roderick Impey Murchison amongst others, in the early part of the nineteenth century.

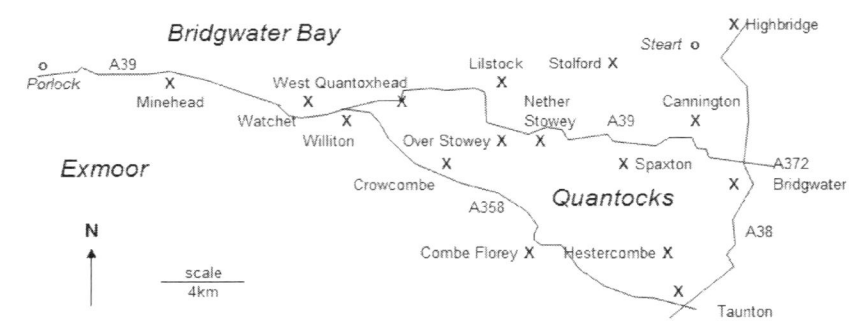

Figure 3.1 Location of some of the places mentioned in the text from the Quantocks, Exmoor, and Bridgwater Bay areas.

The rocks that comprise the Quantock Hills are mainly of Devonian age, that is they are approximately 350 million years old. The oldest of these Devonian rocks are known as the **Hangman Grits** and underlie the highest parts of the Quantocks. These rocks vary greatly from slate to conglomerate (a rock made up of pebbles), and commonly display ripple-marks which are typical of a shallow water environment. Fossils occur in some parts of the Hangman Grits and include seashells and corals. The Hangman Grits occur in a rectangular shaped outcrop from West Quantoxhead to Crowcombe then across towards Over Stowey and back up to Holford.

Plate 3.1 A view across the low-lying Somerset Levels and Bridgwater Bay as seen from the Summit of the Quantocks.

The Hangman Grits have been subdivided in the Quantocks into three divisions. The lowest, called the Little Quantock Beds, outcrop in a small area south of Crowcombe and comprise distinctive brown to grey **siltstones** and slates. Rare fossils may be found, mainly seashells, which indicate a marine environment at this time. The Little Quantock Beds are rather enigmatic and it is not clear whether they should be considered part of the Hangman Grits at all. The next division is known as the **Triscombe Beds** and occurs between Triscombe and West Quantoxhead, it is about 500m thick and mainly comprises green sandstone and mudstones, although some conglomerates do occur. The uppermost division of the Hangman Grits is the **Hodders Combe Beds** and consists of thick beds of sandstone and conglomerate. It reaches a thickness of about 300m, but yields very few fossils, mainly plant fragments.

After the deposition of the Hangman Grits, the sea appears to have become deeper as slates and limestone of the so-called **Ilfracombe Beds** were deposited. The limestones are rich in fossils, mainly corals, but other kinds of fossils do occur as well. It outcrops in the area around Gib Hill and north and east towards Over Stowey and Spaxton respectively.

It is quite a complex group of rocks comprising a number of different beds. The lower beds are called the **Avill** Group and comprise slates, siltstones, and sandstones which contain some fossils, and also a green volcanic deposit known as the **Cockercombe Tuff**, which indicates that active volcanoes must have been close by. Above this group the **Cutcombe Slate** occurs which consists of grey to brown slates and a distinctive layer of limestone called the **Rodhuish Limestone** which is very rich in fossil corals. Another limestone occurs above the slates, called the **Roadwater Limestone**, which again is rich in fossils, but this time of all kinds of marine life. The uppermost beds are referred to as the **Leighland Beds** and are mainly slates and sandstones with some layers of fossiliferous limestones.

Overlying the Ilfracombe Beds is a group of monotonous grey to maroon slates known as the **Morte Slates**, some 1200m thick and devoid of fossils. These are then superseded by the earliest Carboniferous rocks of the region, the **Rodway Beds**, which comprise siltstones and sandstones which again lack fossils, outcropping east of Nether Stowey. Later Carboniferous rocks are seen at Cannington Park. These are limestones and yield abundant marine fossils.

Throughout the Devonian and Carboniferous, the rocks of the Quantock Hills and surrounding area (Table 3.1) record the progressive deepening of an ancient sea. However, this was brought to an end at the end of the Carboniferous when to the south of Britain (Britain at the time was in the southern hemisphere), two drifting continental plates collided to form a mountain range extending in a line from Brittany to Poland. This period is assigned to the Variscan Orogeny and uplifted the whole of southwest Britain above sea-level. It also contorted the rocks, producing faults and folds of which there are many examples in the Quantocks.

Formations	Subdivisions	Rock types	Period
Rodway Beds		siltstones and sandstones	Carboniferous
Morte Slates		maroon-coloured slates	Devonian
Ilfracombe Beds	Leighland Beds	slates and sandstones with some limestones	
	Roadwater Limestone	fossiliferous limestone	
	Cutcombe Slate	grey to brown slates and the Rodhuish Limestone	
	Avill Group	slates, siltstones and sandstones (includes Cockercombe Tuff)	
Hangman Grits	Hodders Combe Beds	sandstone and conglomerate	
	Triscombe Beds	green sandstones and mudstones	
	Little Quantock Beds	brown to grey siltstones and slates	

Table 3.1 The general geological succession of the Quantocks.

The Quantocks region was uplifted to form part of a new land-mass, exposing the rocks to weathering and agents of erosion. Materials eroded from the uplifted Quantocks were deposited on the flanks of the hills, during the Permian period, which contain within them eroded angular fragments of Devonian and Carboniferous rocks held together by a fine red sediment. The angular nature of the fragments identifies this rock-type as a breccia and suggests that they have not travelled far from where they were eroded, and certainly have not been transported by water, which usually wears down rock fragments to become pebbles. The red colour of the finer material indicates that these rocks were laid down in an arid continental climate.

The area was still above sea-level in the following geological period, the Triassic. The so-called New Red Sandstone was deposited during that time, and covers the old land surface around the Quantocks. The environment of these times continued to be hot, and arid desert-like conditions prevailed, but towards the end of the Triassic, sea-levels began to rise again, flooding the Triassic deserts with warm shallow seas. These marine deposits comprise alternating layers of limestones and shales, which gives rise to the local quarryman's name for these rocks, Lias, and they are attributed to the Jurassic period. The Lias may be seen well-exposed on the coast at Watchet, for example, not too far from the Quantocks. There are no younger rocks found in the Quantocks.

However, during the ice age the Quantocks were never covered by ice sheets or glaciers (although some evidence for limited **glacial** activity on Exmoor has been put forward), but probably had seasonal snowfields. At any rate, the ground would have been permanently frozen, and an environment very similar to the present day arctic tundra would have persisted until about ten thousand years ago. Products of ice age erosion form a deposit known as head, and this head can be found in extensive spreads on the Quantocks. Also, because of the relatively large volume of meltwater released from the soil during the annual spring thaw, many small drainage valleys were eroded, many of which may still be seen, but typically lack modern streams (Plate 3.2). These streamless valleys are called dry valleys and are a clear indication that a tundra-like environment once persisted in the area.

Plate 3.2 The Quantocks incised with many small dry valleys.

It is the variety of rocks and the story they tell that has made the Quantock Hills a classic area for geological study, from tropical coral seas and volcanoes, to continental collision, mountain formation, deserts and ice ages. These studies have been underway for nearly two hundred years and are likely to continue for many generations of earth scientists to come.

Case Study 3.1: Tundra Environments

During the last ice age Somerset was to a large extent spared the devastation caused by the advance of glaciers and ice sheets. They came quite close, but stopped just short in the Bristol Channel. Nevertheless, Somerset was gripped by cold icy conditions, which persisted for many thousands of years.

On the northern shore of the Bristol Channel, sediments deposited by glaciers are commonly encountered, especially around Cardiff. These glacial sediments are called till and are characterised by individual sediment grains of all different sizes being mixed up together.

On the southern shore of the Bristol Channel, there are no occurrences of glacial till associated with the last ice age. This area would have formed a zone directly in front of the glaciers, and although it lacked glacial ice it would have been extremely cold. The environment at this time would have been very similar to the present day arctic tundra. It would have been a very open landscape, but with some patches of trees, such as birch, pine, willow and juniper, which provided sufficient herbage to support a number of animals, such as mammoth, woolly rhinoceros, horse, bison, reindeer, red deer, and giant Irish deer.

The ground would have been permanently frozen in this environment, as it is in the present day tundra - a condition called permafrost. Under these conditions, water would generally be rare, so little erosion of rock by water would occur, and consequently little sediment would be deposited. The commonest kind of weathering which does occur in this type of environment, is a process known as freeze-thaw. This occurs where temperatures become high enough during the day to melt a little water, which then seeps into cracks and joints in exposed rocks. Then as temperatures fall during the following night the water refreezes and, upon freezing, expand, and by doing so prise the rock apart. This also happens on a seasonal (summer-winter) scale. Rocks may then fall off vertical faces to accumulate at the base of the cliff as a deposit of angular rock fragments called **scree**. Scree deposits in Somerset aren't very common, but some examples can be seen, such as in Burrington Combe and Ebbor Gorge on the Mendips.

In areas where a soil had formed, the top few centimetres of the soil would have thawed out during the arctic summer. The meltwater produced from this thaw would have been unable to seep into the underlying frozen ground, and so a surface layer of soil and water would have been produced. On sloping ground, this surface layer, lubricated by the meltwater, would readily have slipped over the underlying frozen ground. In this way the surface soil flows downslope by a process known as **solifluction**, and the soil, complete with bits of angular rock, would usually end up being deposited at the foot of the slope. This type of deposit is known as head (a type of colluvium), and characteristically comprises a mixture of soil and angular rock fragments, and occurs on the flanks of hills or in valley bottoms.

In Somerset, head deposits can be found at the foot of the Quantocks, Mendips, and in certain valleys (Plate 3.3). Where the soil flow doesn't reach the foot of a hill or a valley, head may accumulate on *the hillslope as distinct terraces. Indeed, the north facing slope of Black Down on the Mendips has such suites of these terraces. As one walks up to Beacon Batch on the summit of Black Down, one will encounter short steep climbs, alternating with longer and gentler slopes; these are known as the scarp and tread terrace components respectively.*

Plate 3.3 Head deposits at the foot of Brean Down. The deposits at the base of the section have yielded animal remains.

Even though glaciers didn't quite reach Somerset during the last ice age, it was equally cold. These conditions gave rise to their own distinctive forms of erosion and deposition which still survive today, attesting to the severity of the environment at that time.

PART 2 - SOMERSET'S HILLS AND VALLEYS

Chapter 4. Somerset's Ancient Hills and Valleys

Following the earth movements that created the upland regions of Somerset, there was a period of erosion which began carving into the newly uplifted landscape. This erosion which took place in the Permian and Triassic geological periods, created deep valleys and also left upstanding hills, the legacy of which is still apparent today in the Somerset Landscape. Two such deep valleys are the Yeo and Axe Valleys on the north and south side of the Mendips respectively. These two valleys are explored below as examples of these ancient Somerset landforms. The deposition of Triassic rocks into these valleys is then investigated around Barrow Gurney (see Figure 2.1).

4.1 Hills of the Axe Valley

It might seem slightly contradictory that hills may be found within a valley bottom; however, in Somerset this is not an unusual sight. The formation of hills in the Axe Valley, as we see them today, represents millions of years of development, and is a testament to the magnitude of environmental changes that Somerset has experienced.

Plate 4.1 The Axe Valley with Cheddar Reservoir and Nyland Hill in the distance.

The Axe Valley is generally a southeast-northwest trending feature, bounded to the north by the Mendips, an upland area underlain by Carboniferous rocks, and to the south by the Isle of Wedmore, an area of low-lying hills composed of Jurassic rocks (Plate 4.1). There are a number of hills in the Axe Valley, which rise steeply and abruptly out of the Axe Valley Levels. The main hills are Nyland Hill near Draycott, Lodge Hill, Windmill Hill and Chalcroft Hill near Westbury-sub-Mendip, and Knowle Hill near Henton and Yarley (Plate 4.2). The summits of these hills range in height from the 40m high Chalcroft Hill to over 76m on top of Nyland Hill. The latter affords splendid views of the Axe Valley, and has been exploited by film makers for its excellent views of Glastonbury Tor to the east.

Plate 4.2 Knowle Hill in the Axe Valley.

The hills of the Axe Valley would have once been part of the Mendips, and record a previously southern limit of the upland. Geologically speaking, these hills are known as **inliers**, because they comprise older rocks surrounded by younger rocks. The core of most of these hills comprises hard rocks belonging to the Carboniferous Limestone Series, very similar to the rocks found on the Mendips. These sedimentary rocks were deposited some 360 million years ago in the Carboniferous Period, when this part of Somerset lay at the bottom of a fairly shallow warm tropical sea, teeming with life, such as corals, bivalves, and brachiopods (types of sea-shell).

Towards the end of the Carboniferous Period upheavals in the earth were brought about by the collision of southern and northern European continental plates in what has been termed the Variscan Orogeny which caused these rocks to be uplifted out of the sea to become dry land. During the following geological periods, the Permian and Triassic, rivers cut down through the Carboniferous rocks of the Mendips to produce deep valleys, and indeed the Axe Valley is one of these. These valleys were then completely or partially infilled by **Permo-Triassic** sediments, arid continental deposits comprising large quantities of eroded Carboniferous rocks. These Permo-Triassic rocks can be seen flanking most of the Mendips, and surrounding the lower slopes of the Axe Valley hills. Therefore, one can be fairly certain that these hills were first formed over 245 million years ago.

Following their formation, many of these hills may have subsequently been submerged by the rising sea during the Jurassic period. Although there is no direct evidence on the Axe Valley hills, one only has to consider that nearby Glastonbury Tor, which rises to 158m at its summit (over twice as high as Nyland Hill), is entirely composed of Jurassic aged marine rocks, which means the sea-level in this area during the Jurassic period would certainly have drowned the Axe Valley hills and undoubtedly covered them with Jurassic sediment. Indeed, the Isle of Wedmore, which forms the southern boundary to the Axe Valley, as already noted, is underlain by Jurassic rocks. Thus, the hills which were sculptured during the Triassic were smothered by Jurassic deposits.

This situation probably prevailed until quite late in the geological record. Approximately 2 million years ago the ice age began and although this area was probably never extensively glaciated, it would undoubtedly have resembled the modern day Arctic tundra, with perhaps local ice caps on the higher ground, such as the Mendips. Much snow and ice would have accumulated during ice age winters, most of which would melt during ice age springs. Such a large amount of meltwater would have charged the ice age river Axe, giving it immense powers of erosion and ability to transport eroded material. Furthermore, it is known that during the ice age sea-levels were well over 100m lower than they are today, so the coastline would have been well out in the Irish (Celtic) Sea. Under these circumstances, rivers like the river Axe (Plate 4.3) would have attempted to erode and cut their valleys deeply in an attempt to reach sea-level.

Plate 4.3 A tributary of the River Axe at the foot of Knowle Hill in the Axe Valley.

Under these conditions, it is likely that the river Axe removed all the Jurassic rocks from its path, leaving only the older Carboniferous rocks as hills. This preferential erosion is brought about by the difference in hardness of the two rock types; Jurassic rocks are very soft when compared with the hard Carboniferous Limestone. In this way, these ancient hills were exhumed from beneath the cover of Jurassic rocks.

As a consequence of this process during the ice age, many of the river valleys of the Somerset Levels are extremely deep, some extending to 40m below present sea-level; however, these rock-cut valleys are difficult to appreciate today because they have been infilled by later sediments to produce the Somerset Levels. The ice age wasn't one long cold period, it was a series of glacial episodes interspersed with warmer **interglacial** stages. Sea-levels were low during the glacials, but were high during interglacials. Therefore, when sea-levels were high, the rivers were effectively ponded back which allowed sediment to be deposited in the valleys, and in this way the valleys were infilled with sediment. The Axe Valley Levels are underlain by soft sediments of marine origin attributed to the present interglacial (yes, we can almost certainly expect another glacial episode to occur in the future) brought about by the post-glacial rise in sea-level, but in other places sediments of earlier interglacials can be found, such as the **Burtle Beds** of the Brue Valley.

The Axe Valley Levels have been reclaimed and drained, a practice which was initiated during Roman times, with Roman salterns, field systems, settlements, and other earthworks being clearly visible. This indicates that reclamation was extremely successful here, because in other parts of the Somerset Levels reclaimed land was often re-inundated by floodwater. Little attention has been given by archaeologists to the Axe Valley hills, but they undoubtedly served as dry islands for habitation from stone age times onwards.

The history of the development of the Axe Valley and its hills is complex, spanning 300 million years of geological time, a period of continental collision, high and low sea-levels, episodes of deposition and erosion, and climatic changes from an arid Saharan-type environment to ice age glaciations. It is a beautiful part of the Somerset landscape, best seen from the green summits of the hills themselves.

4.2 The Yeo Valley

The River Yeo rises near Compton Martin and flows westwards into, and out of, Blagdon Lake, to become what is shown by the Ordnance Survey as the Congresbury Yeo. The valley the River Yeo occupies has been a valley for millions of years, and beneath the valley floor the rocks provide a record of events that lead up to the present day (Plate 4.4).

Plate 4.4 The Yeo Valley with Blagdon Reservoir in the distance.

The Yeo Valley is basically an east to west orientated valley, bordered to the south by the Mendips and to the north by the high ground upon which Bristol Airport has been built. There are very few exposures of rocks in the Yeo Valley, but by investigating the surrounding uplands one can get an idea of what rock types underlie the Yeo Valley.

The oldest rocks in the area are the Old Red Sandstone, deposited mainly in the Devonian geological period, which can be seen outcropping on Black Down on the Mendips, and are also well-exposed in the cliffs at Portishead. These rocks were deposited some 375 million years ago, in a setting dominated by deltas fed by rivers flowing in from the north. Fossil fish can sometimes be found, as can some fossils of the earliest waterside plants.

The Old Red Sandstone was succeeded by the invasion of the sea into the area during the Carboniferous period. The sea was warm and tropical with abundant sealife, the remains of which can be found as fossils, such as seashells and corals. The main rock type formed at this time has become known as the Carboniferous Limestone, and good examples can be seen in Burrington Combe.

Both the Old Red Sandstone and the Carboniferous Limestone can be seen in the Mendips, and here both sets of rocks appear to be inclined to the north. In the hills to the north of the Yeo Valley, the Carboniferous Limestone is again inclined, but this time to the south. This suggests that there is a downward pointing bend or fold in the rocks directly underlying the Yeo Valley (Figure 4.1). This type of folded structure is known as a syncline.

In a syncline, the rocks at its edge are always older than the rocks in the middle, and in the middle of the Yeo Valley syncline, there occurs a group of rocks that are not exposed in the surrounding uplands. These rocks are also Carboniferous in age and are known as the Coal Measures. They were deposited in a tropical environment, when the area was a swamp, perhaps similar to present-day mangroves. Many animals lived there and some can be found fossilised; however, it is the fossil plants which are really important in the Coal Measures, as it is the layers of fossil plants which form coal, and give the Carboniferous its name.

Following the deposition of the Coal Measures, the area was subjected to the effects of a collision of continents, the Variscan Orogeny. A consequence of this collision, was the upheaval of the land. The area was raised above sea-level at the end of the Carboniferous, and was subject to erosion by water and wind. Thus, an erosional surface, or ancient landscape, was carved into these rocks, producing the first Yeo Valley. Lying on top of this landscape, and infilling and obscuring this early Yeo Valley from view, is a layer of rocks deposited during the Triassic period, when the environment was rather like the present-day semi-arid regions of the Mediterranean. These Triassic rocks are known as the **Mercia Mudstone Group** (part of the New Red Sandstone), and because they were deposited after the Variscan Orogeny, they escaped folding, and so still lie more or less flat, as they were originally deposited in the early Yeo Valley some 240 million years ago.

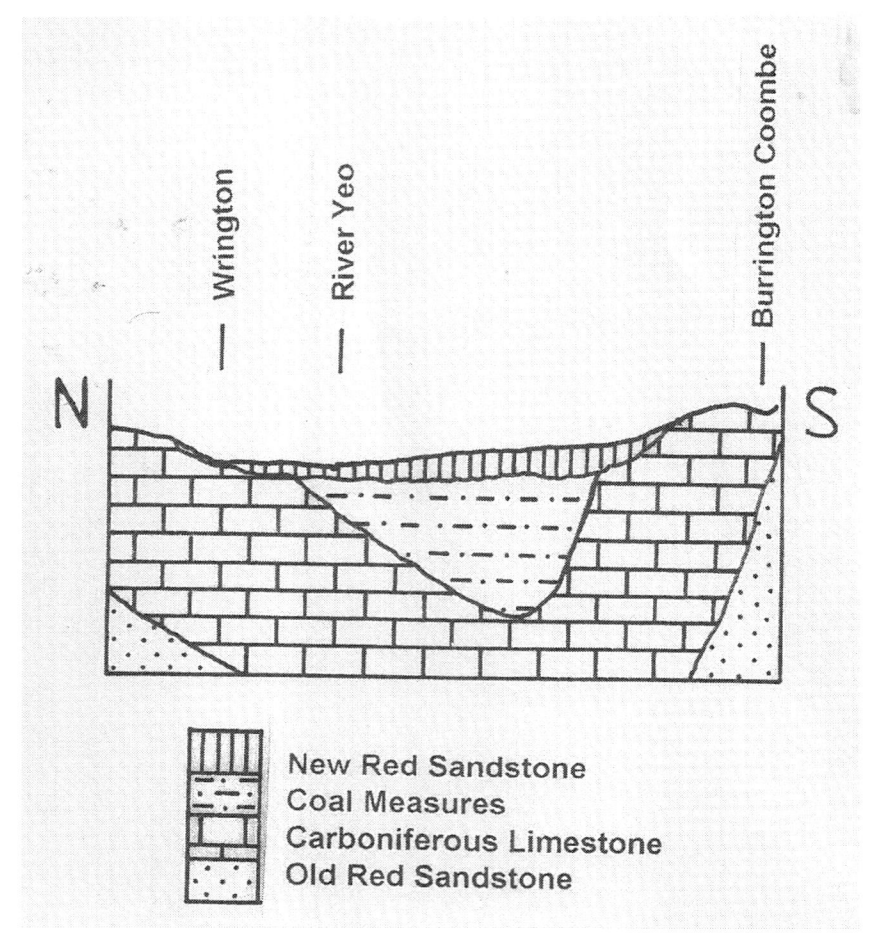

Figure 4.1 The geological structure of the Yeo Valley - a good example of a syncline.

As we have seen, the Yeo Valley has had a long and eventful geological history, a history of changing environments, from tropical seas to arid deserts, and periods of mountain building with associated erosion, which formed an ancient landscape which is now hidden, but was undoubtedly a precursor to the Yeo Valley we see today.

4.3 Barrow Gurney

If one was to walk from Flax Bourton to Dundry, one would pass over rocks that represent an environmental transition from land to sea. This event took place gradually over 200 million years ago, at a time when dinosaurs roamed the Earth, an event that has determined the type of rock we see in this area.

The rocks we see around Flax Bourton are called, the Mercia Mudstone Group of the Triassic geological period. They were deposited in a semi-arid environment, where rainfall was low and intermittent, with most occurring as deluges creating flash floods. This terrestrial environment, inhabited by dinosaurs, occurred when the British Isles were located much further south than at present, around the area now occupied (not unsurprisingly) by North Africa. As a result of continental drift the British Isles have since drifted northwards to their present position.

By the end of the Triassic, sea-levels began to rise to flood the continent. This rise in sea-level is used to define the top of the Triassic, and the subsequent geological period is known as the Jurassic. So in this area, the Jurassic period is heralded by the appearance of marine deposits within the geological record. This transition from continental to marine conditions is well preserved in the rocks around Barrow Gurney. The first rock type encountered above the Mercia Mudstone Group is a pebbly deposit that contains abundant fossils of seashells, such as oysters, which lived on the ancient seashore. Therefore, this rock has been interpreted to represent a beach that migrated over the Triassic surface as the sea-levels rose.

This beach or 'littoral' rock type passes upwards into slightly deeper water sediments, although could still be classed as a shallow sea. This porcellaneous limestone is pale in colour and is known as the **White Lias**. Again, it contains shallow sea fossils, but lacks pebbles. Overlying the White Lias comes the **Blue Lias**, which comprise grey/blue coloured limestone layers interspersed with dark shale layers. It is the rock type that possibly gives the Lias its name, from a corruption of the word 'layers', describing the obvious limestone-shale alternations. Alternatively, the name may be a derivation of Limestone and shale. Fossils are again common in these rocks, and for the first time **ammonites** may be found. In fact, the first occurrence of ammonites in the Lias of Somerset is used as the international definition of the Triassic-Jurassic Boundary.

The Lias extends some way up the lower slopes of Dundry Hill, but most of the rocks towards the summit of Dundry Hill belong to the Inferior Oolite. This is a middle Jurassic rock, deposited when sea-levels had stabilised in a warm shallow water environment similar to the Mediterranean today. Fossils can be found, but this rock is most famous for the excellent building stone that used to be quarried from quarries on Dundry Hill - the Dundry Freestone. This is no longer quarried and the quarries now lie derelict.

Exposures may be found in the area around Barrow Gurney where one may see the various rocks that depict a major environmental event that affected this area in the geological past, an event that has had an influence on the landscape we see today.

Chapter 5. Somerset's Younger Hills and Vales

Following the terrestrial conditions which prevailed during the Triassic, Somerset was once again inundated by the sea which deposited great thicknesses of sediment during the Jurassic and Cretaceous periods, which subsequently started to be eroded during a phase of gentle uplift during the Tertiary period. The Tertiary earth movements left the Jurassic and Cretaceous rocks largely unaltered, and great tracts of these horizontally bedded rocks outcrop throughout Somerset. Examples of areas where these Mesozoic rocks occur, and where subsequent landforms have been created upon them, include the area around Evercreech, Cadbury Hill, Pennard Hill, Ham Hill, Doulting, Dundry Hill, and chalkland areas (see Figures 2.1 and 5.1).

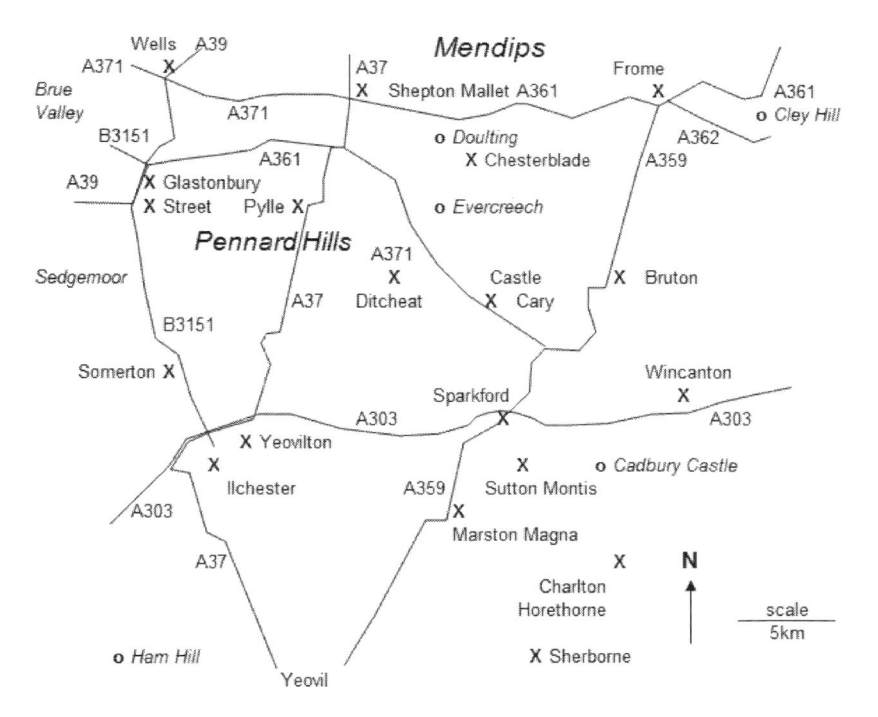

Figure 5.1 Location of some of the places mentioned in the text in the southeast Somerset area.

Somerset Landscapes

5.1 Evercreech Vale

Evercreech village is situated south of Shepton Mallet and to the north of Castle Cary. On entering the village, one is struck by the appealing light grey to buff coloured stone with which many of the buildings have been constructed. If one looks closely at the walls of the buildings, many fossils can be seen, including aesthetically pleasing ammonites. The presence of particular types of ammonites and the character of the rock indicates, to the trained eye, that the stone used in the construction of the houses is Lias (Plate 5.1).

Plate 5.1 Characteristic limestone/shale layers of the Lias exposed in a disused quarry west of Evercreech.

The Lias is a well-known series of sedimentary rocks that are widespread throughout southwest Britain, in a belt extending from Gloucestershire and the Glamorgan coast, southward through Somerset to Dorset, where it is well-exposed at Watchet and Lyme Regis. There are two informal divisions of Lias which I will refer to, a lower White Lias and an upper Blue Lias. The White Lias and the lowermost part of the Blue Lias are of Triassic age, but most of the Blue Lias belongs to the Jurassic geological period. Thus, the Lias straddles the Triassic-Jurassic boundary, some 208 million years old.

The Blue Lias is characterised by layers of limestone alternating with soft shale. Some of the limestone layers are hard and durable, easily breaking into blocks convenient for building purposes. Locally, the Blue Lias outcrops throughout the Evercreech district and can be seen exposed in patches around the village. Old overgrown disused quarries can be found exhibiting the alternating Blue Lias layers. It was quarried as a building stone and for its use in the manufacture of lime. The Blue Lias around Evercreech lies more or less horizontally and hasn't been affected by earth movements to any degree since it was first deposited. This attribute has played an important role in shaping the landscape around Evercreech, which is predominantly flat to undulating (Plate 5.2).

Plate 5.2 Flat landscape west of Evercreech, produced by horizontally-bedded Lias.

To the east of Evercreech, the land rises abruptly from the flat-lying Lower Lias lowland, to form an impressive escarpment comprised of younger Middle and Upper Lias, and Middle Jurassic rocks. Creech Hill (Plate 5.3) forms the southernmost promontory of this escarpment visible from Evercreech, although the winding course of the escarpment actually extends from near Milborne Port in the south to Frome in the north. This escarpment was favoured by early settlers to the Evercreech district, such as Iron Age people, including the Romans, with the latter building a temple on the summit of Creech Hill. A second scarp slope can be seen to the west of Evercreech, where the Middle and Upper Lias of the Pennard Hills occurs. The Carboniferous Limestone of the Mendips forms the northern limit of the Lias in the district.

Plate 5.3 A view of the undulating landscape between Evercreech and distant Creech Hill.

The depositional environment of the Lias was that of a warm shallow tropical sea teeming with life. Many different types of fossils can be found in the Lias, being best preserved in the hard limestone layers. The fossils that may be encountered commonly include ammonites, bivalves, and brachiopods, and also the remains of giant swimming reptiles, such as Ichthyosaurs and Pleisiosaurs.

The types of ammonites found in Evercreech walls generally belong to two species: *Schlotheimia angulata* and *Arietites bucklandi*, the latter being far more commonly seen than the former. Ammonites are some of the most useful fossils known to geologists because they allow rocks to be dated with a high degree of precision. The two species encountered in Evercreech characterise two adjacent time zones within the Lower Jurassic and point to a **Hettangian** to **Sinemurian** age (200 million years ago).

There is some controversy over the way in which the layers of the Lias formed; one school of thought suggests that as the sediment was deposited, it became compacted with the result that the pressure caused calcium carbonate in the sediment to migrate and to concentrate into layers which later became hard limestone. The other school of thought maintains that the alternating sediments were deposited under an alternating environment, such as a rising and falling sea-level, or a variation in the sediment input into the sea from the land.

If the environmental idea is correct, then one would expect the same number of limestone to shale alternations to occur at different places throughout southwest Britain, because one would expect the environment in the region to alternate at the same time. However, detailed measuring of exposures has shown that no two places has the same number of alternations, thus supporting the compaction theory. Yet where studies have been carried out of the microfossils contained in the shale and limestone layers, it appears that the organisms were responding to changing environmental conditions (principally sea level changes), therefore supporting the environmental theory.

Most geologists now accept the microfossil evidence, suggesting that the deposition of the Lias is environmentally controlled. But if it is, what is causing the environment to change so frequently and with such regularity? It is now widely accepted that any environmental changes that occur on a regular basis, such as the 20 or so ice ages and interglacials that the earth has experienced in the last 2 million years, are the result of changes in the shape of earth's orbit around the sun, and to the wobble and tilt of the earth's axis. All these astronomical changes occur on a regular basis and are called **Milankovitch Cycles**, named after the Yugoslavian scientist who first noted them. Thus, aspects of the outcrops of Lias that we see around Evercreech are perhaps an expression of the workings of the universe.

5.2 Cadbury Castle

The remains of Cadbury Castle are sited on a distinctive hill which looks down to the west onto the low ground of the Brue and Yeo Valleys, and to the north over undulating hills, but to the south and east a steep line of hills rise to form the skyline. The site of Cadbury Castle is a natural vantage point with excellent views in all directions (Plate 5.4), making it an easily defended home for the early inhabitants of this part of Somerset. But how was such an ideal defensive site created, to what extent did the populace of Cadbury Castle alter the natural landform from its original state and, moreover, how is nature attempting to reclaim that which has been altered?

Plate 5.4 The western flank of Cadbury Hill rising steeply from the surrounding countryside.

The site of Cadbury Castle is located southwest of South Cadbury just south of the A303 (T). There is a visitors' car park south of South Cadbury Church, and the Castle may be reached by way of Arthur's Lane, leading up the hill from the Church. There is an information board just off the road, which is worth reading. Arthur's Lane is a reminder that Cadbury Castle has long been considered a candidate for the site of Camelot.

Our story, however, begins some 200 million years ago, during the Jurassic Period. At this time in the earth's history Somerset lay at the bottom of a warm subtropical sea, teeming with life of all kinds. As the animals died they fell on to the seafloor and were covered in sediment to become fossils. This sediment hardened to form shale, limestone and sandstone, and the fossils they yield include exquisitely preserved ammonites, sea-shells, and uncommonly, marine reptiles. The site of Cadbury Castle is composed of Lias, and as one walks up Arthur's Lane to the summit, more resistant pieces of limestone and sandstone can be found, sometimes with fossils attached.

The sea occupied this part of Somerset more or less continuously from the beginning of the Jurassic up until about 65 million years ago, when worldwide sea-levels fell. Once land was created, agents of weathering (water, wind and ice) began to erode the rocks that were deposited whilst Somerset was under the sea. Erosion proceeded by peeling off each layer of rock, and in this part of Somerset, erosion has recently (geologically speaking) been removing the upper layers of the Lias and the overlying limestone called the Inferior Oolite, a hard limestone comprised of little egg-shaped grains called ooliths. The high ground around Corton Beacon to the south of South Cadbury is still capped by the Inferior Oolite.

At some unknown point in the past, the site of Cadbury Castle possibly formed the corner of a bent (L-shape) escarpment, with one arm of the escarpment running west to east from Cadbury Castle to Compton Pauncefoot, and the other arm extending southward from Cadbury Castle, along what is now known as Corton Ridge. At present, the steep escarpment is set back to the east and south of Cadbury Castle, and rises to Pen Hill at the northern end and Corton Beacon to the south. The steep face is known as a scarp slope, but from the top the ground descends more gently to the east towards Charlton Horethorne, and this is known as a dip slope.

It appears that streams started to cut into the escarpment perpendicular to one another, one flowing to the north and the other to the west, to where the villages of South Cadbury (Plate 5.5) and Sutton

Montis respectively, now stand. Eventually, as the two streams eroded back into the escarpment, the site of Cadbury Castle would have become distanced from the scarp slope and more and more isolated, until only a single ridge connected the hill with the main escarpment at Charwell Field. At this point, there would have still been two streams running either side of Cadbury Castle, one with its headwaters at Pen Hill and the others at Whitecombe. However, the connecting ridge must have broken down at some point, and the Pen Hill stream switched valleys to join the Whitecombe stream. This left the valley leading to South Cadbury dry, as at the present day; however, a stream still issues from South Cadbury to become the River Cam. Thus, with the ridge eroded as well, the site of Cadbury Castle became a completely isolated hill.

Plate 5.5 A view of South Cadbury from Cadbury Castle.

Early settlers to the area would have found the landforms very similar to today, but the landscape would have been tree-clad. Some of the earliest evidence for human occupation of the site is of Neolithic or New Stone Age. A series of pits were dug into the bedrock and subsequently filled. Flint arrowheads, pottery, and plant and animal remains have been found, including human bones, and date from about 4,755 years ago. The site continued to be occupied through the Bronze

Age, and may possibly have been a hill-fort with defensive structures. However, it was in the Iron Age that the most prominent ramparts and ditches were constructed to defend the site, and these can be seen as concentric rings encircling the hill like a series of steps (Plate 5.6). There were at least six different ramparts built, the first around 400 BC and the last at approximately 40 AD, shortly before the site was attacked by the Romans in 60 AD (around the same time as the Boudiccan rebellion) who massacred the Durotrigian tribe that lived there. The Romans maintained a military and religious presence at Cadbury Castle throughout their reign, and even after the Romans left, it is thought to have been a military stronghold for an important Dark Age chief - possibly Arthur!

Plate 5.6 Impressive ramparts of Cadbury Castle.

Although man-made, these ramparts can be considered landforms, and they dominate the landscape of Cadbury Castle Hill. So the hill itself is certainly natural, but the features which adorn its surface are of human origin; however, if one looks closely at the slopes of the ramparts, evidence can be found to suggest nature is attempting to reclaim altered ground. Soil and rocks are moving down from the ramparts to settle in the intervening ditches, and in this way will eventually leave a uniform slope. There are number of ways in which this is being achieved: rabbits burrow into the ramparts, bringing soil and rocks to the surface which rain washes down into the ditches; small landslips occur and can suddenly move quite a considerable amount of soil and rocks downslope; and at a much slower pace, soil can move gradually down the ramparts under gravity by a process known as soil creep, producing distinctive terracettes which are used by two and four legged pedestrians as pathways, thus speeding up the process.

Cadbury Castle Hill is a rare site that is fascinating for its combination of landscape, history, and legend. But perhaps best of all, on a clear day, the summit of Cadbury Castle Hill affords spectacular views for miles around - the very reason why people lived there in the past and one of the reasons why people will continue to visit it for many years to come.

5.3 Pennard Hill

Pennard Hill is an elongate upland area situated in the heart of east-central Somerset, lying directly between the village of Evercreech and Glastonbury. It is about 8 km long and 3 km wide and rises to a height at its summit on its eastern end of 143 m (approximately 470 ft). Its geology is exclusively Jurassic in age, which is around 200 million years old, so by no means new. However, the British Geological Survey has recently re-investigated part of Pennard Hill, including the area to the east around Evercreech and Batcombe, and has adopted newly formalised names for the rock-types that are found there (Plate 5.7).

Pennard Hill is contained on the British Geological Surveys' geological map of the Glastonbury district and on the currently available map (published 1973) is shown to be underlain by rocks referred to as Lias, with Lower, Middle and Upper Lias present. This simple division of the Lias in Somerset is coarse and cumbersome when compared to the geology of the Dorset coast from Lyme Regis eastward, where the same Lower Lias is subdivided (from oldest to youngest) into the now legendary (well at least amongst amateur fossil collectors) Blue Lias, **Shales with Beef**, **Black Ven Marls**, **Belemnite Marls** and **Green Ammonite Beds**. The Middle and Upper Lias on the Dorset coast is similarly subdivided. One obvious reason for the difference in perceived geological diversity between the Dorset coast and inland east Somerset is the lack of exposure; most of east Somerset is agricultural, whereas the Dorset coast is actively eroding, so maintaining ever-fresh exposures.

Plate 5.7 A view from east Pennard Hill of Pylle and the Bath and West Showground beyond.

In the early 1990s, the British Geological Survey re-visited the area around east Pennard Hill, Evercreech and Batcombe and conducted new investigations into subtleties in the local geology. Out of necessity, geological investigation in countryside like east Somerset involves crawling along ditches filled with stagnant water, wading through streams, and fighting through brambles, nettles and rubbish in long abandoned quarries for the grubbiest of rock exposures (take this from one who knows). This is a far cry from the way Hollywood portrays geologists at work, such as Pierce Brosnan tackling a volcano in Dantes Peak, or Sam Neill unearthing dinosaur bones in Jurassic Park.

From its field work however, the British Geological Survey published a technical report in 1993 (number WA/93/89 for those who may be interested) in which they use some subdivisions within the Lower and Middle Lias. These subdivisions have been included on the current geological map for the Wincanton district (published 1996) which covers the area immediately east of Evercreech, including Stoney Stratton,

Milton Clevedon and Batcombe. The names that have been used on the Wincanton map to replace Lower and Middle Lias are (from oldest to youngest): Blue Lias, **Pylle Clay**, **Spargrove Limestone**, **Ditcheat Clay** and **Pennard Sands** (Table 5.1). One of the nicest things about these names I feel is that, with the exception of the Blue Lias, they are all named after local placenames.

The Blue Lias is the lowest of the Lower Lias local subdivisions and is the classic Lias rock-type with alternating blue/grey limestones and black shales. It forms the low ground around Evercreech and can be seen in some old quarries there, although you don't have to fight the brambles because most pre-Second World War houses in Evercreech are built from Blue Lias limestone quarried locally. Indeed, many ammonites have been built into the walls of houses, all specimens that are preserved well enough to be identified belong to the species Arietites bucklandi with its distinctive furrowed keel, a species that is characteristic of the upper layers of the Blue Lias at Lyme Regis.

The Pylle Clay overlies the Blue Lias and is named after its occurrence on the lower slopes of Pennard Hill at the hamlet of Pylle. It comprises dark grey shaly mudstones in a bed up to 20 m thick. Exposures of Pylle Clay are hard to come by, and most occur in the banks of streams that have eroded into the mudstone. In one such situation I have collected superbly preserved ammonites, most no bigger than an old fifty pence piece, and have identified some of them as belonging to the species *Oxynoticeras oxynotum*. This suggests that the Pylle Clay is at least partly equivalent to the Black Ven Marls of Dorset.

The Pylle Clay is overlain by the enchantingly named Spargrove Limestone. Although named after small exposures of it that are found around the hamlet of Spargrove near Evercreech, it is known to extend to Ditcheat on the southeast flank of Pennard Hill. It is described as muddy limestones and thin silty mudstones, and is nowhere greater in thickness than 2.5 m. Yet despite its meagre thickness, the hardness of the limestone means that it resists erosion and so forms prominent ridges running along the hillslopes, and small waterfalls where it crosses streams. It is uncertain what the Dorset equivalent is to the Spargrove Limestone, and it may be that it does not have an equivalent because a well-known erosion event occurs in the Dorset sequence at the most likely horizon, between the Black Ven Marls and the Belemnite Marls.

Jurassic Lias Divisions	Pennard subdivisions (with max. thickness, m)	Ammonite zones	Dorset equivalents (with max. thickness, m)
Middle Lias (part)	Pennard Sand (40m)	*Amaltheus margaritatus*	lower Middle Lias
	Ditcheat Clay (60m)	*Prodactylioceras davoei*	Green Ammonite Beds (32m)
	Spargrove Limestone (2.5m)	*Tragophylloceras ibex* *Uptonia jamesoni* (part)	Belemnite Marls (23m, part)
	Pylle Clay (20m)	*Echioceras raricostatum* *Oxynoticeras oxynotum* *Asteroceras obtusum*	Black Ven Marls (46m) & Belemnite Marls (23m, part)
Lower Lias (part)	?no strata	*Caenisites turneri*	Shales with Beef (21m, part)
	Blue Lias	*Arnioceras semicostatum* *Arietites bucklandi*	Blue Lias (32m) & Shales with Beef (21m, part)

Table 5.1 The geological succession of Pennard Hill and the surrounding area, correlated with the Dorset succession.

Ditcheat Clay succeeds the thin Spargrove Limestone and comprises silty clays and silts. It dwarfs the Spargrove Limestone in being up to 60 m thick in places and is named after its occurrence around the village of Ditcheat. It underlies Pennard Hill up to about two-thirds of its height, and because it is soft it often creates landslips along the sides of Pennard Hill. It is most likely equivalent to the Belemnite Marls and perhaps the Green Ammonite Beds of Dorset, although the true relationship will depend on further collection of ammonites for comparison.

Finally, forming the resistant summit of Pennard Hill are the sandy beds of the Pennard Sands, which give rise to well-draining sandy soils ideal for the local orchards (Plate 5.8). The Pennard Sands are quite fossiliferous in places, up to 40 m thick, and during Saxon times were quarried like Ham Stone of Ham Hill to the south. They are well exposed in sunken lanes leading down from the summit to West Pennard and fossils can be collected here (Plate 5.9). It is because of the resistance of the Pennard Sands that Pennard Hill and Glastonbury Tor are still standing, for where it has been removed by erosion, the soft sediment of the underlying Ditcheat and Pylle Clays has been stripped away to create the Brue Valley to the south and the course of Whitelake to the north.

Plate 5.8 An apple orchard on the summit of Pennard Hill, making the most of the relatively free-draining soils of the Pennard Sands.

Plate 5.9 Cottles Lane, at the western end of Pennard Hill, sunk into the fairly resistant Pennard Sands.

The revisit of the British Geological Survey to east Somerset has no doubt re-invigorated interest in the Jurassic geology of Somerset in recent years. It has diversified the geological landscape and in doing so has stimulated discussion and raised new questions to be answered by further investigation. This is an excellent example of how the discipline and study of geology, and science in general, has evolved, ever striving to refine further our initially broad understanding of the world.

5.4 Ham Hill

Ham Hill rises above Stoke sub Hamdon in south Somerset. It has long been famous for the unusual stone which characterises the area, known as the Ham Hill Stone. As well as being of value because of its aesthetic and building properties, Ham Hill Stone is also of great geological interest.

Ham Hill is often described as a well-defined plateau. This description arises from the steep climb up the side of the hill, and because of the relatively flat expanse of the 84 hectare hill-top. The summit of the hill is capped by a relatively resistant rock-type - the Ham

Hill Stone. The underlying rocks, known as the Yeovil Sands, are softer and are more easily eroded. Therefore, once the overlying Ham Hill Stone has been removed, erosion of the Yeovil Sands proceeds rapidly, resulting in the steep sided hills we see at Ham Hill.

Both the Ham Hill Stone and the Yeovil Sands are sedimentary rocks (Plate 5.10) that were deposited during the Jurassic geological period, which began some 208 million years ago. More precisely, they are both considered to be Lower Jurassic in age, and have been attributed to the geological formation known as the Upper Lias. The Yeovil Sands occur below the Ham Hill Stone, and so are slightly older.

The Yeovil Sands occur over a wide area in south Somerset, from Sherborne in the east through Yeovil to Crewkerne in the southwest. They comprise layers of fine-grained sand, crumbly sandstone, and sandy limestones, and together are about 40m thick at Ham Hill. Fossils are not common in the Yeovil Sands, particularly in the lower sections, but fossils such as ammonites can be found in the higher parts. The Ham Hill Stone on the other hand is a shelly limestone which is almost completely made up of fossils, though, nearly all the fossils have been badly broken. It is only when the rock has been sliced and viewed through a microscope that the fragments of shell can be seen clearly. However, some whole identifiable fossil specimens of ammonites and brachiopods (a type of seashell) can be found on rare occasions. The Ham Hill Stone is about 25m thick on Ham Hill, but it is much thinner on other hills nearby, such as at Gawlers Hill, Chiselborough Hill and Chinnock Hill to the south. Apart from these occurrences, the Ham Hill Stone is found nowhere else.

Despite the lack of fossils in the Yeovil Sands, there is abundant evidence that the sand has been thoroughly churned up by the burrowing of soft bodied animals which never became fossils themselves. Normally only animals with a hard shell are likely to become preserved as fossils, as soft body parts decay away shortly after death. This burrowing is likely to have been performed by worm-like creatures either living on or within the sediment, rather like in modern day sand flat environments, where lugworms churn up the sand. Thus, the Yeovil Sands has been interpreted as being deposited in a sandy shore environment.

Plate 5.10 A disused quarry face on the summit of Ham Hill exhibiting sedimentary structures, such as primary and cross-bedding.

Interestingly, there is an equivalent sand layer to the Yeovil Sands in north Somerset known as the **Midford Sands**, which occurs around Bath. This sand layer extends southward to Yeovil where it becomes the

Yeovil Sands, and continues southward to the Dorset coast, where it is known as the **Bridport Sands**. From the fossils contained in the Sands at each locality, it has been possible to establish the age of each of them. Generally, the Midford Sands in the north are the oldest, whilst the Bridport Sands in the south are the youngest, with the Yeovil Sands being intermediate in age. The fact the Sands become younger from north to south suggests that the sandy shore was migrating southward during the period in which deposition was taking place, acting rather like a modern day offshore sand-bar.

As this sand-bar migrated southward, the amount of sand available decreased, because it was being deposited during migration, and by the time the sand-bar reached the Dorset coast the sand supply was nearly exhausted. Channels formed through the sand-bar, excavated by powerful tidal currents, linking the open sea to the south with the water body behind the sand-bar to the north. One such channel, and a major channel at that, broke through the sand-bar in a line northeast of Crewkerne. Abundant shell material accumulated in this channel, deposited by the tidal currents moving through the channel in a northeasterly direction. This flow direction is established by looking at bedding in the rock faces, where sediment accumulated on the sloping lee side of sea-bed ripples, producing what is known as cross-bedding. Thus, in this ancient coastline, the flood tide was undoubtedly much stronger than the ebb tide. These accumulations hardened to become the shelly limestones we now call Ham Hill Stone.

The area was inundated by the sea following the breakdown of the sand-bar, and a shallow marine limestone was deposited at the start of the Middle Jurassic, known as the Inferior Oolite, which can be seen outcropping from Sherborne, through Yeovil and Crewkerne, to the Dorset coast. Although not found on Ham Hill itself, patches of Inferior Oolite can be found above the Ham Hill Stone at West and Middle Chinnock, and on Chiselborough Hill. It comprises small egg-shaped sediment grains or ooids, which formed in a warm subtropical sea where gentle currents rolled around the ooids, successively coating them in limescale. Fossils are usually abundant and well-preserved in the Inferior Oolite, in particular ammonites are exquisitely preserved and highly sought after.

The uniqueness and aesthetic value of the Ham Hill Stone has long been acknowledged, and the stone has been quarried at least since Roman times (Plate 5.11). However, the hill top has been occupied for considerably longer than that. Evidence of Neolithic (new stone age) people on Ham Hill is abundant, with many flint scrapers, axes, and arrowheads being found over the years, in addition to pottery fragments and some stones that originally came from Cornwall. Bronze age artefacts have also been found, which suggest that the hill may have been occupied and defended well before the onset of the Iron age, when it became occupied by people of the Durotrigian tribe. The Durotrigians were opposed to Roman occupation, but a systematic programme of pacification by the Romans in Durotrigian territory led to the development of a major Romano-British settlement on Ham Hill, complete with a Roman villa. There is little evidence to suggest what replaced the Romans on Ham Hill following their departure, but it is certain that the Saxon invaders continued to quarry the stone.

Plate 5.11 The clear impact of quarrying on Ham Hill.

Ham Hill is a key site for unravelling the complex geological history and geographical setting of Somerset during the Jurassic period. The Ham Hill Stone, restricted to Ham Hill and a few hills nearby, is also of value as a building stone. The resistance of Ham Hill Stone to erosion influenced the development of the steep-sided plateau that is now Ham Hill, producing an ideal site for early Neolithic people to dwell, and for later populations to exploit, in terms of quarrying. Today one is free to explore Ham Hill to the full, investigating the rock faces in the disused quarries, and walking amongst the spoil heaps produced by two thousand years of quarrying.

5.5 Doulting

The village of Doulting and its famous stone are unconformable for a number of reasons. First, it is a village of the Mendips, and like many other Mendip villages it has gained its livelihood from quarrying. Yet the stone it extracts is not the almost ubiquitous Carboniferous Limestone of most of the other Mendip quarries, but is a much younger limestone from the Jurassic geological period. Secondly, the rocks of Doulting themselves exhibit a geological phenomenon known as an **unconformity**.

Doulting lies just east of Shepton Mallet on the A361 toward Frome. It is a charming village with numerous old buildings built from the local stone. There are many attractions in the village, such as St. Aldhelm's church, a Medieval barn, and a spring known as St. Aldhelm's Well which is the source of the River Sheppey. There is parking near the church and some local facilities are available. The rocks discussed below are exposed in the working quarry east of the Doulting to Chelynch road, although permission for visitors is needed from the quarry manager (Plate 5.12). Older quarries may be seen from footpaths in woods east of the active quarries, many of which are overgrown, although some excellent rock faces still exist (Plate 5.13). A once famous railway cutting also exists to the south of Doulting.

The Mendip Hills are famous for the Carboniferous Limestone that has been quarried from their flanks since at least Roman times, but the stone quarried at Doulting is much younger in age, from the Jurassic geological period. In fact, Doulting Stone is part of a particular Jurassic rock type known as the Inferior Oolite and was deposited during the middle of the Jurassic. Oolite is a form of limestone that comprises unusual grains of lime. The grains are almost perfectly round and spherical, and are known as ooids, because they are essentially egg-shaped. Oolites are formed in shallow, warm, lime-rich seas where waves are able to move and roll sediment on the sea floor. In these warm seas lime precipitates out of the water, rather like lime in a kettle precipitates out to form limescale. However, the limescale in the sea coats itself to the rolling sediment grains and builds up concentric layers of lime to form an ooid. Ooids are being formed today in places like the Caribbean, where the water is shallow, warm and lime-rich.

Plate 5.12 Quarried Doulting Stone in the working quarry between Doulting and Chelynch.

At Doulting the Inferior Oolite rests directly upon Carboniferous Limestone and so the contact between the two rock types represents a period in time for which we have no geological record. The actual contact is not exposed at Doulting today, but a similar situation occurs further east along the southern flank of the Mendips at Vallis Vale near Frome. Here horizontal beds of Inferior Oolite overlie inclined Carboniferous Limestone, the upper surface of which has been eroded and bored into by animals prior to the deposition of the Inferior Oolite. When there is a clear break between the deposition of one sedimentary rock and another, the upper bed is said to lie unconformably upon the lower bed, and the contact between the two is known as an unconformity.

Plate 5.13 A disused Doulting Stone quarry in woods near Doulting.

This type of hiatus in the deposition of the sediments which now form these rocks, has always filled me with great wonderment. Millions of years passed between the deposition of the Carboniferous Limestone and the Inferior Oolite, the intervening upper Carboniferous (Coal Measures), Permian, Triassic, and Lower Jurassic (Lias) rocks are all missing. Thus, this break, the unconformity, represents millions of years, during which Britain drifted north from the equator to the mid-latitudes, and dinosaurs evolved to dominate the land. When confronted with an unconformity, I often find myself imagining being able to become two-dimensional and slot myself into the unconformity, and then to stretch the time period to the boundaries imposed by the underlying and overlying layers of rock.

The Inferior Oolite is so-called because it is the lower division of the Middle Jurassic oolites, the upper one being referred to as the Great Oolite Series (Table 5.2). The Inferior Oolite itself is divided up by geologists, based on subtleties of sediment type and fossil composition, into the Lower, Middle and Upper Inferior Oolite. The Lower and Middle Inferior Oolite are missing from Doulting, and the Upper Inferior Oolite lies either directly onto Carboniferous Limestone or upon Lias in a few places. The Upper Inferior Oolite is the main economically important rock that was quarried at Doulting. Architecturally, Doulting Stone is one of the best known of the Upper Inferior Oolite rocks, as it has been used in a number of local buildings, including Wells Cathedral and Glastonbury Abbey (Plate 5.14). It is a limestone of uniform composition and with very few lines of weakness, which makes it ideal for dressing and stone work.

Epoch	Division	Members	Rock-type	Comments
Bathonian	Fuller's Earth			
Bajocian (Inferior Oolite)	Upper Inferior Oolite	*Anabacia* Limestone	iron-stained oolitic limestone	contains *Anabacia* corals
		Doulting Stone	oolitic limestone	shallow water, exhibits cross-bedding
		Doulting Ragstone	oolitic limestone	shallow water environment
		Doulting Conglomerate	pebbly limestone	fossil beach deposit
(unconformity)				
Dinantian	Carboniferous Limestone			

Table 5.2 The general geological succession at Doulting.

Plate 5.14 Glastonbury Abbey: a well-known monument constructed using Doulting Stone.

This famous Doulting Stone is actually only one of four Upper Inferior Oolite beds that occur at Doulting. The lowest and oldest is a bed of limestone containing pebbles known as the Doulting Conglomerate. It contains fossil worm tubes and borings, and fossil brachiopods (a type of sea-shell). This bed is thought to represent a shore deposit and is only about 50cm thick. As the Jurassic sea rose up the flanks of the Mendips it laid down these shoreline deposits which were later covered over. The Doulting Conglomerate is superseded by the **Doulting Ragstone**, an oolitic limestone of some economic importance which comprises the abundant remains of sea-lilies or crinoids and sea-shells. Some hard and bored layers within the Ragstone may indicate temporary disturbance of the sediment whilst it was being deposited. This may be due to deepening water and the effect of waves passing over the sea-bed at the time.

The Doulting Stone follows the Ragstone but, although it is very similar in that it is an oolite containing abundant crinoid remains, it is more consistent, lacking the hard layers of the Ragstone. However, it does exhibit the work of ocean currents in that the sediment is deposited as ripples on the sea-bed and these can be clearly seen in the cut blocks. The ripples are at an angle to the horizontal bedding of the oolite, which has given rise to the name of this type of sediment structure, which is cross-bedding (Plate 5.15). The combined thickness of the Ragstone and Doulting Stone is nearly 14m which contributes to their economic importance, with up to 1000 tonnes quarried annually at Doulting. Overlying these beds is the **Anabacia** Limestone, a brownish iron-stained oolite characterised by button-like fossil *Anabacia* corals. This is the youngest bed of Upper Inferior Oolite at Doulting, and the sequence is overlain by less than 2m of Fuller's Earth, which is seldom exposed.

Plate 5.15 Cross-bedding in Doulting Stone at Doulting.

For a small village, Doulting boasts numerous attractions for the historian and geologist. For the geologist, it is rare these days to find an active quarry working Inferior Oolite and so fresh exposures and clean rock specimens are worth the visit. The Carboniferous Limestone, the skeleton of the Mendips, is here unconformably overlain by a thin skin of Jurassic rocks which record the rising of a long-gone sea, complete with a fossil beach and evidence for the work of waves and currents. However, you don't have to visit Doulting to see Doulting Stone, there is plenty of it on show at Wells Cathedral, Glastonbury Abbey, old railway stations to Cheddar..........such is its aesthetic significance.

5.6 Dundry Hill

The rocks that cap Dundry Hill are part of an extensive and important geological formation called the Inferior Oolite, which stretches in a band from the Dorset coast in the south to the Cotswolds in the north. Yet these rocks at Dundry Hill are now isolated from the main outcrop of the Inferior Oolite, which survives to the east, where it forms the hills between Radstock to Bath. Thus, the Dundry rocks indicate that substantial erosion has occurred in this part of north Somerset, stripping away the Oolite and exposing older rocks below. This has left Dundry Hill as a geological island of younger rocks surrounded by lower and older rocks, a geological structure called an **outlier**.

Dundry Hill is situated close to the southern outskirts of Bristol and rises to a height of 233m. There are a number of footpaths which traverse the summit, most offering splendid views of the surrounding countryside, Bristol, and further afield, including Wales on a clear day (Plate 5.16). At Dundry village, car parking is available near the church and local services are on hand. A footpath leads from the car park and heads westward across open farmland.

Plate 5.16 View of the countryside of Dundry Hill towards Bristol in the distance.

The Inferior Oolite that caps Dundry Hill was deposited during the middle of the Jurassic period of geological time. Oolite is a form of limestone that comprises unusual grains of lime. The grains are almost perfectly round and spherical, and are known as ooids, because they are essentially egg-shaped. Oolites are formed in shallow, warm, lime-rich seas where waves are able to move and roll sediment on the sea floor. In these warm seas lime precipitates out of the water, rather like lime in a kettle precipitates out to form limescale. However, the limescale in the sea coats itself to the rolling sediment grains and builds up concentric layers of lime to form an ooid. Ooids are being formed today in places like the Caribbean, where the water is shallow, warm and lime-rich.

Fossils are very common in oolite, although their state of preservation can be sometimes variable. The Inferior Oolite is very famous amongst geologists for the exquisitely preserved ammonites that it yields (eg Plate 5.17). Many ammonites have been found on Dundry Hill over the last two centuries, many of which can now be seen in museums. Many of these specimens were collected when quarrying was active on Dundry Hill and new specimens of fossils were unearthed daily. However, all the quarries are now overgrown, with the rock faces covered by vegetation or lichen, and fossils more difficult to come by. The only quarrying that persists today is the extraction of rubble from the old quarry spoil heaps, which is used by local farmers for hardcore. However, where these old spoil heaps have been opened up, fossils may be found amongst the rubble. Besides ammonites, bivalve and gastropod molluscs are common, as are brachiopods (another type of seashell), and corals.

The Inferior Oolite is so-called because it is the lower division of the Middle Jurassic oolites, the upper one being referred to as the Great Oolite Series. The Inferior Oolite itself is divided up by geologists, based on subtleties of sediment type and fossil composition, into the Lower, Middle and Upper Inferior Oolite. The Lower and Middle Inferior Oolite are represented by very thin beds at Dundry of three and two metres respectively. It is unfortunate that these are seldom exposed today, as they yielded a great many fossils. The fossils that have been recovered are very similar to those found in Dorset, but unlike those found in contemporary rocks in the Cotswolds, suggesting that somewhere between Dundry and the Cotswolds there was a change in the environment which influenced the types of animals living in the two areas.

Plate 5.17 *Stephanoceras humphriesianum*, an ammonite species of biostratigraphical importance in the Inferior Oolite (although it has not yet been recorded from Dundry).

The Upper Inferior Oolite is much thicker and includes the main economically important rocks that were quarried at Dundry. Architecturally, the Dundry Freestone is perhaps the best known of the Upper Inferior Oolite rocks (Plate 5.18). Dundry quarry was owned by the church, and the stone has been used in a number of local religious buildings, including the church St. Mary Redcliffe in Bristol. It is a limestone of uniform composition and with very few lines of weakness, which made it ideal for dressing and stone work. Curiously, it is a local rock-type only, with no equivalent workable limestones of the same age in either Dorset or the Cotswolds. Indeed, its occurrence on Dundry Hill itself is somewhat variable, occurring as a bed nearly ten metres thick at the western end of the hill, but barely over one metre thick at the eastern end. Other Upper Inferior Oolite beds include coral beds, grits, and other limestones, all of which contain fossils.

Plate 5.18 An old quarry face of Dundry Freestone on Dundry Hill.

The routes up to the summit of Dundry Hill are steep, but at the top a rather flat and plateau-like landscape is seen. This is because the rocks that make up Dundry Hill are all lying horizontally. This may not seem unusual, as one would expect that all layers of sediment are deposited horizontally under gravity. This is true, however, many rocks once formed may be affected by earth movements which can compress and tilt them. In north Somerset most rocks that are encountered are inclined in this way, but Dundry is different. That is because the Jurassic post-dates the period of active earth movement in this region, so that whilst the older Carboniferous rocks of the Mendips, for example, are highly contorted, the Jurassic rocks are in relatively the same position as they were when they were originally deposited, some 180 million years ago.

The hard limestones of the Inferior Oolite only occur at the very summit of Dundry Hill, and overly the weaker alternating limestone and shales of the Lias. The Lias is soft and can be easily eroded away from underneath the Inferior Oolite. Where this happens the oolite is undermined and collapses. In this way, the slopes of Dundry Hill have become littered with the debris from toppled Inferior Oolite to comprise what geologists call foundered strata.

The record of human occupation of Dundry Hill is somewhat sketchy. There is evidence for Iron Age activity with the discovery of pottery dated to 200-300 BC and attributed to the Dobunni tribe of north Somerset. This is not unexpected given the close proximity of Dundry Hill to the major Iron Age hillfort of Maes Knoll. Following the Dobunni, it is known that extensive quarrying of the oolite took place at Dundry Hill during the Roman occupation, and has probably continued to some degree ever since.

Dundry Hill is a major landmark in north Somerset and a popular recreational area for the people of Bristol. Not only does it afford spectacular views over city and countryside, but it is crucial in understanding the geological history of southwest Britain from the Cotswolds to Dorset. Also, like other areas of limestone, the alkaline soils provide a suitable environment for downland plants, which on a summers day carpet rough ground around the hilltop and attract large numbers of bees and butterflies, blues, browns and fritillaries.

5.7 Somerset's Chalklands

Chalk is by far the most familiar rock known to the inhabitants of southern Britain. The extent of the Chalk stretches from Kent in the east to Devon in the west, Yorkshire in the north to the Isle of Wight in the south. Where Chalk is exposed at the coast it produces magnificent scenery such as the White Cliffs of Dover and the Needles off the western tip of the Isle of Wight.

Here in Somerset only a few hills in the southern part of the county are underlain by Chalk, especially around Chard, but it also outcrops in Wiltshire, Devon and Dorset close to Somerset's border. It produces well known landscape features, yet despite our familiarity with Chalk, its natural history is not well known.

Chalk is a soft white limestone, technically called a **biomicrite** by geologists. It is almost totally composed of calcite, the mineral form of calcium carbonate, with some mineral impurities such as quartz, iron-pyrites (fools' gold) and a green mineral called **glauconite**. Pure Chalk has over 80 per cent of its composition made up of calcite, while Chalk with less than 80 per cent calcite starts to become impure. A common rock associated with Chalk is flint, which occurs as bands of nodules in the Chalk and, in stark contrast to the pure white Chalk is usually dark in colour. Flint nodules are normally composed of 100 per cent silica, and are thought to form by the leaching of silica out of the Chalk (derived from **sponge spicules** and **radiolarian** microfossils) and concentrating it in a nodular form.

Unlike most sedimentary rocks, which are made up either of clays or sands, Chalk is composed almost entirely of the skeletons of tiny organisms. These micro-organisms include plate-like algae called **coccoliths** and small chambered animals known as **foraminifera**. By far the largest volume is provided by the coccoliths, which when living floated freely in the surface waters of the sea. Some foraminifera also inhabited the upper waters of the Chalk sea and led a planktonic mode of life, whist others lived on the sea floor and are said to be benthonic (from the Greek work benthos meaning depth of sea).

Larger animals are also occasionally found preserved as fossils in the Chalk. Fossil snails, cockles, mussels, sponges and sea-urchins are commonly found, as are extinct animals such as ammonites and bullet-shaped **belemnites**. All these organisms which make up the Chalk lived in a warm subtropical sea. Although it is difficult to ascertain, it is largely accepted that the Chalk environment was moderately deep and quiet, fluctuating around 100 fathoms.

The deposition of the Chalk in Somerset spans a time period of some 35 million years, the oldest Chalk being nearly 100 million years old and the youngest 65 million. This falls within the upper part of the geological period called the Cretaceous, which is itself named after the Latin word *Creta* which means Chalk. In the beginning of the Cretaceous dinosaurs roamed the land, but by the end of this period they had become extinct.

The Chalk is subdivided into the Lower, Middle, and Upper Chalk, and examples of all three may be seen in Somerset. The Lower Chalk is characterised by a fairly thick succession of white Chalk, which is well-bedded and quite soft. This outcrops west of Chard on Snowdon Hill and yields exquisitely preserved ammonites. The Middle Chalk consists of variable Chalk, some with flints and some without. It occurs between Crewkerne and Chard, at Lady's Down and Cricket St. Thomas. The Upper Chalk is again quite variable and has a limited outcrop in Somerset, being found at only a few localities north of Chard.

The landscape that the Chalk produces is extremely characteristic. Where the Chalk lies level it produces large flat plateaus, however, where the Chalk has been inclined high hills and ridges are formed. In places where the Chalk has been eroded away the termination of the Chalk produces a steep escarpment which rises to a high ridge or **cuesta**, which then dips gently away.

Unfortunately, although Chalk does occur in Somerset, it is of limited extent, not well exposed and difficult to see. Undoubtedly the best exposures of Chalk, and the most dramatic chalkland scenery, occurs in Wiltshire, but very close to Somerset's eastern border. Places like White Sheet Hill near Stourhead east of Bruton, and Cley Hill between Frome and Warminster, are good examples that illustrate the uniqueness of the Chalk and the landscape it creates (Plate 5.19). Steep escarpments, often with step-like terracettes, scrubby vegetation, pale ploughed fields, and numerous archaeological remains are characteristic. Early Britons liked living on the Chalk because it is porous and allows water to percolate downwards, leaving the surface dry and preferable to the damp marshy lowlands, and afforded good views (Plate 5.20). Also, tools could easily be made locally because of the abundance of flint. Cley Hill is a particularly good example as it is managed by the National Trust with ample parking, good footpaths, and the Chalk changes from Lower Chalk to Upper Chalk, as one traverses to the summit.

Plate 5.19 The Chalk ridge of Cley Hill between Frome and Warminster (note the pale soil in the ploughed fields around the foot of the hill).

Chalk has been economically important and to a limited extent still is. During a drive through the Chalk countryside of Somerset or neighbouring Wiltshire one cannot fail to spot the odd tree-clad hollow in the middle of a field or the roadside Chalk cliff, which represent remnants of an old Chalk quarrying industry in the area. Chalk has many long established uses such as an agricultural lime, as a component in cement, in the chemical and iron industry and, strangely enough, in the paper making industry (where it is used to glaze shiny paper).

Plate 5.20 East Somerset (Frome is in the distance), as seen from the chalky summit of Cley Hill.

The Chalk has a remarkable natural history as it is one of the only rock-types in the world to be composed solely of the remains of fossil organisms. Of this 100 million year old sub-tropical sediment it is the middle and upper divisions that are commercially valuable as they are relatively pure, while the lower Chalk contains clay and tends to be impure. Some quarries and roadside verges exist where Chalk is exposed, and anyone can go and collect Chalk, flints and fossils for themselves.

Chapter 6. Recently Formed Valleys

Although Somerset was not widely glaciated during the Ice Age, with perhaps only a few glacial incursions in northwest Somerset, glaciers and the severe cold had a profound effect on the landscape which is still very evident today. Some of the landforms created during this time are explored below, such as the development of Ebbor and Avon Gorges, Burrington and Blagdon Combes and the diversion of the River Chew (see Figure 2.1).

6.1 Ebbor Gorge

Carved deep into the steep southerly aspect of the Mendip Hills, close to the ancient city of Wells, Wookey, and Westbury-sub-Mendip, Ebbor Gorge offers the ambience typical of a British woodland coupled with dramatic scenery characteristic of several Somerset gorges. The rocks which comprise the cliffed walls of the gorge record the environments and events that occurred here many millions of years ago. The gorge itself, sculpted from these rocks, indicates relatively recent geological activity (Plate 6.1).

Ebbor Gorge is most easily accessed along the road from Wookey Hole, which itself is a turning off the A371 from Wells to Cheddar. The area is managed by the National Trust and possesses a car park with ample space, a picnic area, information boards and nature trails. There are many aspects of natural history that may be observed at close hand here, but geologists note that hammers are not permitted.

Most of the rocks at Ebbor Gorge were deposited during the beginning of the Carboniferous Period and are assigned to the Dinantian Series of rock some 350 million years old. The rocks themselves are mainly various types of limestone collectively known in this part of Britain as Carboniferous Limestone, and there are a number of distinct layers of this limestone exposed here.

The oldest layers are known as the Burrington Oolite and can be seen exposed in small cuttings near the Ebbor Gorge Nature Reserve car park. Oolite limestones are made up of very small spheres of lime known as ooids. Fossils may be recovered (mainly corals and brachiopods) but are often seen to be abraded.

Plate 6.1 The Carboniferous Limestone cliffs of Ebbor Gorge.

The next variation in the sequence is called the Clifton Down Limestone and can be seen exposed in Ebbor Gorge itself. It comprises dark grey to black muddy limestones with occasional beds of oolite. Again, it is fossiliferous and fossil corals and brachiopods may be found here. Within the gorge the Clifton Down Limestone has been weathered to produce vast spreads of scree (Plate 6.2).

Plate 6.2 A well-developed scree slope in Ebbor Gorge (note the figures for scale).

However, between the car park and Ebbor Gorge two successively younger beds occur. These are the Hotwells Limestone and Quartzitic Sandstones. Neither is well exposed and yet their presence, sandwiched between the Burrington Oolite and Clifton Down Limestone, both of which are older, poses a problem. In fact their occurrence here is caused by a geological structure called a fault, which is a linear fracture in the rocks, here running northwest-southeast. This has meant that the rocks northeast of the fault (Clifton Down Limestone, Hotwells Limestone, to Quartzitic Sandstone sequence) have been thrust under the older Burrington Oolite on the southwest side. Indeed this fault is an important local geological structure and has been called the Ebbor Thrust.

A thrust fault is produced when a block of rock is compressed. Under compression rock may deform like plastic to produce geological structures called folds, or if it is too brittle to fold, it may fracture to produce a fault. In a thrust fault the block that is thrust up and over an underlying block is called the hanging wall, whilst the block underneath is called the footwall. So in the case of the Ebbor Thrust, the Burrington Oolite belongs to the hanging wall, which has been thrust over the footwall of Clifton Down and Hotwells Limestone, and Quartzitic Sandstone.

This thrusting is the result of the Variscan Orogeny. This collision of continents compressed the rocks forcing great slabs of rock to ride north-eastwards over others. The Ebbor Thrust is an important geological structure which is accessible and easily studied.

Following uplift associated with the Variscan Orogeny, erosion of this ancient landscape occurred. Geographically, Somerset was located just north of the equator at the time, so the climate was semi-arid, but with occasional catastrophic flash-floods. These environmental conditions led to the erosion of the Carboniferous Limestone and the products of this erosion were deposited at the foot of the newly created Mendips. This deposit contains large numbers of grey Carboniferous Limestone fragments cemented in a red-brown matrix. The reddish colour is attributable to the fact that deposition of this rock occurred in air, allowing iron in the sediment to oxidise. This attractive rock is called the **Dolomitic Conglomerate**, dolomite being a common mineral found in them. However, this is somewhat of a misnomer, as in some places it contains no dolomite, and is more often a breccia rather than a conglomerate. Nevertheless, it has been quarried and used extensively in local buildings, including Temple Meads railway station in Bristol.

Ebbor Gorge may have begun to be excavated during this arid time of uplift, but the features we see today are so fresh that more recent erosional processes are likely to be responsible for the landform. It is thought that local environmental conditions associated with the ice age are responsible for the development of the gorge. During this time the ground would have been frozen, a condition called permafrost. This prevented water, derived from seasonal snow melts, escaping underground into the limestone (as water does today on the Mendips). Also, the Quartzitic Sandstone is impermeable. Thus, regardless of bedrock type, water was forced to form torrential streams and rivers flowing over the land surface. It is only through surface flow that erosion capable of gorge formation can take place. Some have even considered that a small local ice cap existed on the Mendips which was drained by the many Mendip gorges, such as Ebbor and Cheddar Gorges, and Burrington Combe. It is ironic that such gorges, which must have experienced incredibly high levels of energy, now lie relatively quiet, empty, and devoid of the water that created them, truly relict features, fossilised in time.

Ebbor Gorge affords excellent exposures of Carboniferous Limestone. Although it must be studied passively, it is nonetheless rewarding, especially the evidence for the Ebbor Thrust. In addition, the gorge is located in a wonderful setting, ideal for a family day out, and I find it particularly inspiring to stand in the bottom of the gorge and try to imagine the great torrents of icy water, that some 18,000 years ago, must have flowed down between the cliffs every spring!

6.2 Burrington Combe

Burrington Combe on the north face of the Mendips is an impressive landform; a valley cut steeply into the hard Carboniferous Limestone. But how was it formed? Valleys are normally formed in one of two ways; first, by the action of rivers, and secondly, by glaciers.

It is widely acknowledged that glaciers probably didn't occur on the Mendips during the last ice age, and besides, glacial valleys are characterised by broad bottom U-shaped valleys and not the narrow steep sided valley that is now Burrington Combe. The only other possibility then is that it is a result of river erosion. However, Burrington river is conspicuous by its absence. So how did Burrington Combe form?

The answer probably lies with the Carboniferous Limestone. This hard limestone has many cracks and joints running through it, which make it permeable to water, and so any water attempting to flow over it will usually vanish into the rock. Once in the rock, water can dissolve the limestone to form caves. It has been suggested that this may have happened in the past, with a large cave being formed, eroded by an underground river which grew too big to support its own ceiling. The ceiling then collapsed to form Burrington Combe. But if this were true, one would expect to find large boulders of limestone strewn along Burrington Combe, as remnants of the ceiling. Such boulders aren't present, so it is unlikely that this theory is correct.

Alternatively, it may have been eroded by a surface river which for some reason was prevented from disappearing underground. During the last ice age the environment of the Mendips would have been similar to the present day arctic tundra, and the ground would have frozen for most of the year, a phenomenon called permafrost. Water would have frozen in the cracks within the Carboniferous Limestone, sealing them and rendering the limestone impermeable. Seasonal melting of small ice caps, thought to have existed on the Mendips during the last ice age, would have had to flow overland, and because the spring melts were probably torrential, there would have been adequate river energy to erode steeply into the Mendips to form Burrington Combe (Plate 6.3).

Although the theory of permafrost preventing water seeping underground is the most likely, there is a third theory which warrants consideration. It is a well-known that many of the world's most impressive **karst** (limestone weathered) features occur in wet tropical climates, and thus the limestone caves of Borneo and gorges of Thailand, are amongst the largest of their kind. This is because in these regions, the very high rainfall dissolves more limestone. Therefore, by analogy, we cannot rule out the possibility that Burrington Combe formed during a time in the geological past when the British climate was much wetter and warmer, than it is today. About 120 thousand years ago, there occurred a warm period that separated the last and penultimate glacial stages, and has been called the **Ipswichian Interglacial**. Sediment deposited during this period on other parts of the country have yielded fossils of animals indicative of very warm conditions. For example, fossil lions and elephants have been found under Trafalgar Square, and fossil hippopotami have been found as far north as the Yorkshire Moors. It is possible that the warmth prevalent at that time was associated with wetter conditions, increasing the rate at which the limestone of Burrington Combe was eroded.

Plate 6.3 The steep sides of Burrington Combe. Workers are making safe the rock faces by clearing vegetation and loose rock.

This controversy over the formation of Burrington Combe is typical of the earth sciences, whereby we try to explain natural processes that have been operating over timescales that are often difficult to comprehend. There is always going to be more than one theory, and often more than one theory is right, showing that natural forces can combine and work together to produce a particular landform. It has also been suggested that its origin goes back to the Permo-Triassic, when the combe may have been similar to a **wadi**, with seasonal flash-floods draining the Mendip massif; indeed, there are Permo-Triassic rocks deposited in the combe mouth. Whatever the correct theory, and it is likely to be a combination of them, Burrington Combe has long been considered a scenic wonder of the British Isles and will no doubt continue to attract many visitors, who may ponder on its formation, for many years to come.

6.3 Avon Gorge

The River Avon flows through Bath and on through Bristol to join the Severn Estuary at Avonmouth, yet this present day course may represent a fairly recent and dramatic deviation of the River Avon from previous courses, a deviation that may have formed the Avon Gorge. The gorge itself, being cut into the local rocks, provides excellent exposures which can be studied at a number of localities.

The Avon Gorge is very narrow and has long been an important transport route from Bristol to the Severn Estuary, both for river transport and the A4 road (and its precursors) which runs along the east bank of the river. Because of this it is very busy and often difficult to gain access to the gorge cliffs. However, it is possible in a number of places, often exploited by rock climbers. The safest and perhaps most pleasant way to enjoy the gorge, and to observe its geology and wonder at the landform itself, is to visit the viewpoint at the eastern end of the Clifton suspension bridge (Plate 6.4) and then to take one of the forest walks in the Avon Gorge Nature Reserve on the wooded west bank of the gorge at Abbots Leigh.

The geology exposed in the cliffs on either side of Avon Gorge is the relatively hard Carboniferous Limestone. These sedimentary rocks were laid down when Somerset lay at the bottom of a fairly shallow warm tropical sea, teeming with life such as corals, bivalves and brachiopods (both these are types of seashells), which can be collected from the gorge today as fossils. Locally, the Carboniferous Limestone has been subdivided using subtle changes in the type of limestone and the different fossils that occur at different levels. Names such as Hotwells Limestone, Black Rock Limestone and Clifton Down Limestone will be familiar with anyone who has studied the geology of the gorge or indeed the Carboniferous geology of Somerset in general, including the Mendips.

Plate 6.4 The Clifton Suspension Bridge straddling the Avon Gorge and River.

Piecing together histories of rivers is very difficult, as the very nature of a rivers activity hinders the preservation of a record in the landscape (Plate 6.5). This is because rivers spend half their life in phases actively eroding the landscape (Plate 6.6), obliterating any sediments that may have been deposited in previous quieter times that may have given us a clue to their history.

Plate 6.5 The muddy, tidal River Avon flowing through the incised Avon Gorge.

Plate 6.6 Cliffs eroded from Carboniferous Limestone in the Avon Gorge.

The River Avon is no exception. Some believe that originally the River Avon did not flow westward to the Severn Estuary, but flowed southwards through the heart of southern Britain. It may have existed in a pre-ice age landscape when Britain was possibly tilted to the south. This is possible because during the ice age the great weight of the northern ice sheets in northern England and Scotland depressed the land they occupied, so tilting the British landsurface to the north. At the present, northern Britain is rebounding upwards in response to the melting of the great ice sheets, and as a consequence southern Britain is actively in the process of tilting southwards again - a bit like a seesaw. Therefore, it is not unreasonable to suggest that in pre-ice age times, Britain had a more southerly tilt which would have influenced river courses (this is reflected, to some extent, in the deposition of some Cretaceous rocks in Somerset, Devon and Dorset). After all, the goal of a rivers life is to flow downhill to reach sea-level. Sometime subsequently however, the River Avon was diverted to flow in a westward direction through Bath, most likely as a consequence of ice sheet formation in northern Britain during the early part of the ice age.

It appears that the new route approximately followed the modern valley to the south of Bristol in which Long Ashton, Flax Bourton and Nailsea are situated, perhaps exiting into the sea at Clevedon. Indeed if one follows the modern Avon, there is a very sharp bend at Ashton Gate, where it is easy to imagine the Avon ploughing straight on to Long Ashton. However, if this is the case, what then made the River Avon change course yet again to flow northwards to Avonmouth?

The area that is now the Avon Gorge would have been relatively high ground and an effective barrier to the river. However, during one glacial event during the ice age, it is known that an ice sheet came across the Bristol Channel from Wales and plugged the coastline at least between Clevedon and Weston-super-Mare. All the rivers that flowed westward in this area were dammed and lakes were formed. When the level of the lakes reached the tops of the hills the lake waters started to overspill and cut down through the hill top rocks to form new valleys.

It is precisely this overspilling that is thought to have cut Avon Gorge, as a way of letting the dammed water escape. There may have been a smaller river valley in existence at this time, which was exploited and captured by the diverted Avon. The Avon is not the only river that may have changed course during this time. The River Chew, combined with the River Yeo, originally flowed westward through the Yeo Valley, but the ice damming reversed it to flow to the north to meet the River Avon.

Evidence for glaciation in north Somerset is not restricted to 'overflow' gorges, as they are called, but we do have some sedimentary evidence. The village of Kenn, near Congresbury, has been constructed in a slightly elevated position above the general altitude of the Levels that surround it. The reason why the ground is higher here is because it is underlain by glacial sand and gravel deposits. These glacial deposits were laid down by the ice that invaded the region from across the present Severn Estuary and Bristol Channel. Also, some small local glaciers were thought to exist in the hills surrounding the Vale of Gordano. However, this glacial activity in the region only occurred during one of the many ice ages during the last two million years, a glacial stage referred to as the '**Wolstonian Glaciation**'.

Although our area did not have much in the way of glaciers itself during the majority of the ice ages, it was affected by ice coming from elsewhere during the 'Wolstonian' and this has left a dramatic impression on the modern landscape of the region, an impression which is manifest in the formation of the Avon Gorge.

6.4 Blagdon Combe

Travelling west along the A368 from Blagdon, one passes through a valley variously known as Blagdon Combe or the Rickford Gap. The valley has a flat bottom and has steep sides rising up to about 50m above the bottom. The valley bottom has been given over to pasture, but the steep sides are well-wooded and very impressive (Plate 6.7).

The length of Blagdon Combe is no more than 1km from Combe Lodge at its eastern end to Rickford at the westernmost extreme. The valleys course is meandering with three impressive spurs jutting out into the valley. The A368 road however, strikes a more or less straight path through the meanders, from one spur head to another. Such a meandering valley course is typical of valleys formed by river or fluvial erosion.

Plate 6.7 The green bottom of Blagdon Combe.

The river that eroded Blagdon Combe is conspicuous by its absence. For the majority of its length Blagdon Combe has no watercourse; there is a small stream which begins halfway along the Combe and feeds the small lake above Rickford Mill, but it is certainly not of a size capable of eroding such a major landform.

The outlet from Blagdon Combe joins the Yeo Valley at Rickford, and if one follows the gently sloping (approximate gradient of 1:50) Combe eastward, you will end up at Combe Lodge just west of Blagdon. At Combe Lodge, Blagdon Combe appears to stop, not against hills where the catchment of the river would be expected to be, but in mid-air some 60m above the floor of the Yeo Valley, with an approximate gradient down to it of 1:20. Thus, the river that must have formed Blagdon Combe either had no catchment, or the catchment that it did have has since disappeared through erosion.

So, the origin of Blagdon Combe is extremely problematic at first sight; not only does the Combe not now possess a river capable of eroding it, but there does not appear to be a catchment for a river ever to have existed, and the head of the valley is now stranded 60m up in the air.

Assuming that there never has been a catchment, as seems to be the case, we must envisage water levels in the Yeo Valley east of Blagdon Combe being at least 60m higher than at present; that is, there must have at some time been a lake occupying the eastern Yeo Valley. The water levels west of Blagdon Combe must have at the same time been lower so as to allow water to flow downhill, and thus it is unlikely that there was a lake in the western Yeo Valley. What could have caused a lake to form in one part of the Yeo Valley and not in the other?

Some geologists have long speculated over this problem and it seems that the most likely answer involves changes in the area that occurred during the ice age. Glacier ice occupying the lower Yeo Valley could act like a dam allowing water levels in the upper valley to rise. This ponded water must have found and exploited a pre-existing weakness in the landscape (such as a small stream valley, for example), where Blagdon Combe now is. The water must have flowed to exit west of the ice dam and eroded the channel of Blagdon Combe we now see.

There is no independent evidence for glacial activity in the Yeo Valley, but as already mentioned there are some glacial deposits nearby at Kenn. It is thought that the glaciers were flowing southward from Wales, and blocked the outlets to numerous valleys in north Somerset.

Other geologists however, have suggested that Blagdon Combe is a relict stretch of the River Yeo, when the main valley was at least 60m higher. It is envisaged that Blagdon Combe was abandoned when the River Yeo apparently switched to its present valley, which has since been lowered through erosion by 60m. This is possible, but considering the amount of necessary erosion that has occurred in the Yeo Valley since abandonment, requiring a significant period of time, and the 'fresh' appearance of the steep sides of Blagdon Combe, it is perhaps unlikely.

Case Study 6.1 Dry Valleys

Many valleys in Somerset are unusual. Their origin and development have remained a puzzle for most of the 19th century, for unlike most hill and valley systems found in Britain, many valleys in Somerset currently lack rivers and streams - they are dry valleys (eg Plate 3.2).

Valleys are normally the result of the action of rivers cutting their way down through rock, exploiting weaknesses (such as faults) in an attempt to reach sea level. However, if no river exists in a valley, then how does the valley form? Victorian geographers suggested that such valleys were produced by small glaciers flowing southward during the ice age. However, Somerset's dry valleys do not possess the typical U-shaped cross-section that is characteristic of glacier carved valleys in Scotland, Wales and the Lake District. Instead, the problematic dry valleys possess a V-shaped cross-section, typical of valleys eroded by rivers. Furthermore, it was later discovered that at the height of the ice age glaciers and ice-sheets did not extend south beyond a line that roughly corresponds with the Severn and Thames estuaries, leaving most of Somerset to the south.

Geographers soon realised that if they were going to find a solution to the problem it would come from facts they could gather and collect rather than from hypothetical theories which could never be validated. So attention turned to the only element of the puzzle that could be scientifically studied - the rock into which the valley was carved. Studies on the rocks underlying dry valleys revealed that water could not flow over it. Spaces between the constituent particles in the rock act like a sponge and allow water to pass through the rock (a porous rock), thus making it impossible for water to remain on the surface. Or, similarly, some other rocks are well-jointed and allow water to escape underground through these cracks (a permeable rock). This discovery explained why there were no longer rivers in these valleys, but the problem of how the valleys originally developed remained.

For rock to be eroded away it must offer some resistance, therefore for the porous and permeable rocks to have been eroded by water, in the form of rivers and streams, it must have been temporarily impervious, so allowing water to flow over it. The only way in which these rocks could have been effectively sealed for a short time is if water in the pore spaces or joints became frozen. This condition, known as permafrost, exists today in the Arctic tundra regions and during the ice-age the environment in Somerset would have been very similar.

During ice age winters the rock would have been completely frozen, whereas in summer the top few inches would thaw, but the bulk would remain frozen. In these conditions rivers and streams could exist on these rock-types carrying huge quantities of meltwater from the seasonally melting glaciers and ice-sheets to the north, eroding the frozen rocks away, and thus producing valleys. In this environment Palaeolithic (stone-age) man may have seasonally roamed the treeless hills and plains following herds of elk, bison and mammoth. These herds of large animals would have migrated annually from the warmer climes of southern Europe, crossing with ease the swampy area that is now the English Channel. Ptarmigans, geese and ducks would have nested in the short dwarf-birch scrub, often near one of the numerous small lakes or ponds produced as a result of seasonally thawing soil. Snowy owls would have flown silently over the landscape, hunting small rodents such as lemmings.

All that is left from this period of Somerset's prehistory are some man-made artefacts such as flint arrowheads, axes and spears, and relict dry valleys where raging torrents once flowed. The valleys now lie quiet and empty, the rivers and streams that formed them are now long gone and forgotten, but they had a great influence on Somerset's landscape as we know it today.

6.5 Chew Valley

All rivers have the same goal - to reach the sea. Most rivers take the shortest route to the sea, starting in the hills and flowing seaward, yet here in Somerset the River Chew flows inland a considerable distance before joining the River Avon on its seaward journey. Why does it do this, especially when one considers that only a low ridge separates the Chew from the Yeo Valley and the sea.

The headwaters of the River Chew arise on the north flank of the Mendips near East Harptree and flow northward. From West Harptree one can see westward over a low ridge near Compton Martin, through the Yeo Valley towards the sea. Yet from here the River Chew continues northward (dammed at Chew Stoke to create Chew Valley Lake, Plate 6.8) to Chew Magna, and from there its course veers northeast, inland, to join the River Avon at Keynsham. Why has the River Chew's course developed in this way?

Plate 6.8 Chew Valley Lake: a modern reservoir in an ice age valley.

All rivers want to reach the sea, or more correctly reach sea-level or base-level. They do this by flowing downhill to the sea, and as they do so they also erode their river beds, bringing the entire river closer to sea-level. The steeper the gradient to the sea, the quicker the river will flow and the more erosion will occur. Once erosion occurs the river becomes trapped within its valley, making changing its course very difficult. Unfortunately for rivers, sea-level is not static, it goes up and down, this means the speed of the river, and its rate of erosions fluctuates with sea-level. If sea-level rises, a river will slow down and erosion will lessen.

The development of the present course of the River Chew is thought to be related to the ice age. It is now widely accepted that during the ice age, glacier ice flowed south from Wales and the Severn Valley, and flowed along the Somerset coastline, in some places flowing some distance up pre-existing river valleys. In doing so this ice effectively blocked off the seaward route of many rivers, forcing them to choose new courses.

The way in which this may have happened is fairly straightforward. The ice would have acted like huge dams, ponding up river water within their valleys to form lakes. With a continuous input of water, the lake level would rise, and eventually the lake level would rise above the height of neighbouring hills. Once this had occurred, the lake would start to empty (rather like an overflowing bath), finding a new path to the sea, and as it did so the running water would begin to erode a new channel and valley which would slowly get lower, gradually emptying more and more of the lake.

At the end of the ice age, the ice sheets retreated northwards, removing the ice dams, and releasing the water previously locked up in the lakes. However, the new channel and valley that formed during the ice age may be retained by the post-glacial river. Thus, it is very likely that before the ice age the Yeo and Chew Valleys were one, with just one river flowing west to the sea, and that it was glacier ice that forced the river to find an alternative course to the sea. In this sense, the present River Yeo is a new river flowing in an old valley (note that it is much smaller than the Yeo Valley in which it flows), and the River Chew is the descendent of the original pre-ice age river, now flowing in a new valley of just the right size.

Case Study 6.2 Coal Mining at Pensford

Pensford is situated within the Chew Valley, which is quite steep-sided. The downcutting of the River Chew during and after the ice age has eroded through the Carboniferous rocks of the area, exposing coal-bearing strata of the Somerset coalfield (Plate 6.9).

The Somerset coalfield may be thought of as three distinct parts: the Nettlebridge area, the Radstock and Midsomer Norton area, and the Pensford area situated in the Chew Valley. Coal has been mined in the Somerset coalfield for nearly 2000 years, since the arrival of the Romans. But it wasn't until the advent of the industrial revolution around the seventeenth century that coal mining became one of the main economic activities of the area. Coal extraction in the Somerset coalfield had its heyday in the early part of the twentieth century, between 1900 and 1925, but started to decline soon afterwards. In 1947, when nationalisation of the coal industry occurred, there were only twelve remaining mines working coal in Somerset. By 1958 there were only five working mines left, and ten years later there were only two. The last mine of the Somerset coalfield ceased working and closed in 1973.

Plate 6.9 The undulating landscape around Pensford, with the tip of a coal spoil heap protruding through the trees in the middle distance.

The coal measures of the Somerset coalfield were deposited in the geological period known as the Carboniferous. Indeed, the presence of coal at this time throughout Europe and North America has actually given the period its name, as Carboniferous means carbon-bearing (coal is mainly composed of carbon). Generally, the coal measures are divided up into the Lower and Upper Coal Measures, separated by a barren (non-coal bearing) layer called the Pennant Sandstone. In the Pensford coal basin only the Upper Coal Measures were mined, and here they were further subdivided into an upper Pensford Group and a lower Bromley Group, separated again by a barren layer of red shales.

During the Upper Carboniferous, when the coal measures were deposited, Britain was straddling the equator and experienced a tropical environment. The climate was warm with high rainfall very similar to the conditions currently prevailing in the Amazon rainforest. Indeed the area would have been swathed in tropical rainforest. The main types of plants that existed then are largely extinct now, but resembled present day club-mosses, seed-ferns, and horse-tails, but with one big difference, they were giant forest trees then and not the small plants we see today.

The Pensford spoil tip used to be a good place to find fossil plants, brought up from the mine.

Because of the high amount of rainfall, coupled with the fact that the land wasn't much above a stable sea-level at the time, the ground would have been almost permanently water-logged, so that when the plants died and fell to the ground they could not rot and decompose because of the lack of oxygen. Thus, stagnant swamp conditions prevailed, and one can imagine the stench of hydrogen sulphide that must have filled the forest. A layer of peat would have formed which later, after burial by more peat and sediment, turned to coal.

Because the land at this time was close to sea-level, there were times when the sea invaded the land, and during these marine incursions layers of sand and mud were deposited which separate individual coal seams. Also, sea-level could fall and so lower the water-table. When this happened coal could not form because the ground wasn't water-logged, allowing the plant debris to decompose. At these times, oxygen could react with iron in the sediment to turn it a rust-red colour, and the barren red shales of the Pensford basin are thought to have been formed in this way.

Generally, the coals of the Pensford basin occur in thinner seams than in the Radstock basin, and are considered to be 'dirtier', that is they contain less carbon and more sediment. However, that said, the Pensford Colliery worked the Pensford No. 2 and No. 3 coal seams until its closure in 1958, and so was one of the later collieries to fold.

PART 3 - COASTAL AND LOWLAND SOMERSET

Chapter 7. Somerset's Old Coastlines

Old coastlines are those that are recycled to some extent. Sea levels have fluctuated constantly throughout geological time and during the Quaternary period alone sea levels may have risen and fallen (by as much as 130m) more than twenty times. Thus, some areas may have been coastlines more than once. They are generally characterised by cliffed coasts of hard rock, where relict cliffs and beaches may be found. The coastlines around Porlock and Brean are such old coastlines and have been the land-sea boundary on more than one occasion (see Figures 2.1 and 3.1).

7.1 The Vale of Porlock

Somerset's coastline is varied, but nowhere is it more spectacular than in west Somerset and north Devon. Steep wooded slopes plunge towards the sea, characterising the coastline between Minehead in the east to Ilfracombe in the west. However, there is one notable interruption in the cliff-line, the Vale of Porlock.

The rocks that comprise the cliffs along this coastline are of Devonian age, that is they are approximately 350 million years old. The oldest of these Devonian rocks are known as the Lynton Beds and are found only on the coast around Lynton in Devon, extending southeastward inland to outcrop in Somerset. These beds are characterised by sandstones and mudstones with thin layers of shells. The sediments appear to be well-churned up by animals that used to live on and in the sediment and this suggests that the Lynton Beds were deposited in a shallow sea. The presence of fossil fish also confirms this.

The Lynton Beds were superseded by a group of rocks generally called the Hangman Grits. These beds record a shallowing of the sea, with deltaic and river sediments being deposited. These rocks vary greatly from slate to conglomerate, and commonly display ripple-marks which are typical of flowing water. Fossils, including seashells and corals, occur in the upper part of the Hangman Grits, heralding a deepening of the sea at this time. The Hangman Grits can found in the cliffs between Lynton and Ilfracombe, and also along the entire stretch of coast between Lynton and Minehead.

Some good exposures of Hangman Grits occur along a footpath running atop the cliffs between Porlock Weir and Culbone. Although clad with stunted oak trees, these cliffs are steep and actively eroding. Fast running streams cascade down the slopes, incising into the rock as they flow (Plate 7.1).

Plate 7.1 A fast flowing river issuing into a lagoon behind the gravel barrier at the eastern end of Porlock.

Landslides also occur frequently, with sections of the footpath often closed because of the risk of landslides. The local authority is attempting to monitor the movement of soil downslope and at certain points along the footpath crude instruments have been erected to do just that (Plate 7.2). These comprise a wooden post hammered into the ground, with notches carved into its side, and a pulley at the top. A stake is then driven into the slope to be measured below the post and tied to a piece of wire. The wire is then threaded through the pulley to dangle alongside the post, weighted down by a concrete block. If the soil on the slope should move, the stake will be carried with it, which pulls the wire, lifting the concrete block. The number of notches the concrete block has moved up the post indicates the distance the soil has moved downslope.

Plate 7.2 Instrument used to measure soil movement at Porlock (see text for full explanation).

After the deposition of the Hangman Grits, the sea became deeper, represented by slates and limestone of the so-called Ilfracombe Beds. The limestone is rich in fossils, mainly corals, but types do occur as well. The Ilfracombe Beds can be seen outcropping on the coast around

Ilfracombe. Inland, these lower and middle Devonian rocks pass upwards into upper Devonian rocks, such the Morte Slates, which underlie much of the Brendon Hills. But on the west Somerset coast the next group of rocks to be encountered belong to the Triassic and Jurassic periods. These outcrop within the Vale of Porlock, and from Minehead eastwards in a straight line to Shepton Mallet.

The Vale of Porlock is an ancient valley that was carved into the Hangman Grits during the Triassic. The Triassic rocks here are simply the debris that was eroded from the upstanding Devonian rocks, and the Jurassic rocks record the occupation of the valley by the sea. These Jurassic rocks yield abundant fossils of seashells and ammonites, which indicate it was a warm shallow sea.

Therefore, the Vale of Porlock has been a valley for many millions of years and was inundated by the sea during the Jurassic period. It has also been flooded by the sea in more recent (geological speaking) times. Following the end of the last ice age, about 10,000 years ago, global sea-levels rose as the great ice sheets of the Northern Hemisphere melted. As sea-levels rose, the sea invaded areas that were previously dry land. The Vale of Porlock was flooded at this time and underlying the valley floor near the sea, there are thick deposits of soft marine clay, typically of a blue/grey colour. Within this marine clay, there are a number of layers of peat, including submerged forests (mainly tree stumps), which indicate that sea-levels fluctuated at this time, with periods of sea-level fall as well as rise. Marine clay and submerged forest beds can sometimes be seen on the foreshore at Porlock Weir, especially after storms.

During this time it is thought that Mesolithic (middle stone-age) people inhabited this area, living on the coast during the winter and moving to the uplands in the summer. Indeed, at Hawkcombe Head on Porlock Common, the remains of flint workings have been found, which suggests that a number of family-sized groups hunted here. There is no naturally occurring flint in the area, and it is thought that flint pebbles were imported for working from Barnstaple Bay.

The present day coastline at Porlock Weir is dominated by a storm beach, that is, a beach comprising shingle, pebbles, cobbles and boulders, which can normally only be moved by storm waves. This storm beach stretches for some 4.5 km across the mouth of the Vale of

Porlock. It migrated inland about 4000 years ago, over the top of the marine clays and peat, to its present position. It is primarily made up of Devonian aged rocks derived from cliffs of Hangman Grits and head (Plate 7.3).

Plate 7.3 Blocks of Devonian rocks in a head deposit at the western end of Porlock Weir: a ready source of material for the storm beach.

Somerset Landscapes

Until 1996, this storm beach acted as a barrier, preventing the sea from regularly flooding the valuable agricultural land behind it (Plate 7.4). It also protects the harbour at Porlock Weir, and allows boats to dock in safety. At the same time it provides shelter for **saltmarshes** to develop; a saltmarsh is a delicate marine environment that cannot survive in areas open to severe wave action. However, on 28 October 1996 a storm broke through the barrier, creating a wide breach which allows water to flood the back-barrier land at high-tide (Plate 7.5). The Environment Agency is allowing this to occur under a scheme of managed retreat.

Plate 7.4 An eastward view of the storm beach (gravel barrier) at Porlock.

Plate 7.5 The breach in the storm beach at Porlock, created during a storm on the 28 October 1996.

The seaward face of the beach often comprises a number of distinct ridges. These indicate the direction in which waves travel when they break on the beach. Usually, these ridges are aligned parallel to the shore, which means that the waves were breaking head-on (called swash-aligned ridges); but sometimes these ridges lie diagonally across the beach, and this indicates that the waves were striking the beach at an angle (known as drift-aligned ridges).

Furthermore, the pebbles and cobbles that can be found on the beach are not distributed randomly. They are sorted by the waves during storms, whereby all the beach material is moved up the steep beach face. The large and round cobbles simply roll back down the steep face and accumulate at the foot of the beach, but flat or disc-shaped pebbles can't roll and so tend to get stuck up on the main ridge. Occasionally, if the storm is strong enough, material can be washed over the top of the beach to accumulate on the landward side, and in this way the whole beach can migrate inland. If sea-levels continue to rise in the future as a result of global warming, then under natural conditions the beach could retreat inland. Indeed, the breach formed in 1996 could be a sign that this is happening.

The coastline of west Somerset is an area of dynamic slope and coastal processes which are fascinating to study, but more than that (but perhaps because of that) the area offers beautiful scenery which accommodates the picturesque village of Porlock Weir and the hamlet of Culbone.

7.2 Brean Down

Walking northwards along the long sandy beaches at Berrow and Brean, the view is blocked by a large rugged-looking hill, jutting out into the Bristol Channel. This is Brean Down, a former island that has been colonised by humans for thousands of years.

Brean Down runs almost east-west and is just over 2km long. It separates Berrow Flats to the south from Weston Bay to the north, where the Axe River meets the sea. Brean Down rises dramatically from sea-level to 97m, and from the car park at the north end of Brean Sands the footpath climbs steeply up the southern flank and requires a certain level of fitness or patience to negotiate. But the effort is amply rewarded by the splendid views afforded at the summit. South to the Quantocks and Exmoor, east to the Mendips and Axe Valley (Plate 7.6), north to Weston-super-Mare, and west out into the Bristol Channel and to Wales.

Climbing up the footpath rocky outcrops can be seen (Plate 7.7). This rock is Carboniferous Limestone, the same kind of rock that underlies the Mendips. Thus, Brean Down is often considered as a westward promontory of the Mendip Hills. Indeed, if one stands at the seaward-most point above Howe Rock, the islands of Steep Holm and Flat Holm can be seen, which too comprise Carboniferous Limestone and hint at the long broken connection of the Carboniferous Limestone hills of Somerset and South Wales.

Plate 7.6 A view east along the spine of Brean Down towards the Mendips.

Plate 7.7 A rocky outcrop of Carboniferous Limestone seen from the footpath that ascends the southern flank of Brean Down (note the Somerset Levels and the Axe Valley in the distance).

The Carboniferous Limestone is a hard and durable rock which is fairly resistant to weathering (Plate 7.8) and so often forms upland regions in Britain, such as the Peak District and Pennines. Such a deposit is characteristic of a warm and tropical marine environment, as the calcium carbonate (lime) dissolved in sea water requires relatively high temperatures for it to precipitate out of the water to form a solid which can then become sediment. This interpretation is also backed up by the abundant fossils that can be found in the Carboniferous Limestone at Brean Down and elsewhere. These fossils include corals, seashells, and crinoids. Crinoids are more commonly known as sea-lilies, but are really animals related to sea-urchins; they comprise a long stem constructed with numerous small discs with a central hole (a bit like a small polo-mint) that supports a set of arms which wave around in the currents at the bottom of the sea, trapping and collecting food particles floating by. Upon death the animal breaks up and it is the small discs from the stem that are most often seen as fossils.

Plate 7.8 A view to the west from Brean Down, with Steepholm and Flatholm islands visible in the centre and right distance respectively: evidence for the resistant nature of Carboniferous Limestone.

The cliffs surrounding Brean Down provide excellent exposures for investigating the Carboniferous Limestone and for fossil collecting; however, this can only be done at low tide or on a falling tide. Another feature that can be observed in the cliffs, is the inclined or dipping nature of individual limestone beds. This suggests that following the deposition of the limestone here, the beds were subjected to earth movements which tilted the rock layers. These earth movements are attributed to the Variscan Orogeny.

Following the Variscan Orogeny and the uplift of the area, Brean Down has been surrounded by desert in the Triassic period and subtropical sea again in the Jurassic and possibly Cretaceous periods, but since then has largely remained above sea level. Over the last two million years, the geological period known as the Quaternary, the northern hemisphere has experienced the waxing and waning of ice sheets, flowing south from the Arctic Circles. During this time the environment of Brean Down has alternated between one of very cold tundra-like conditions and a climate very similar to or warmer than that of today.

We have a record for some of this period at Brean Down, because on its southern flank, just seaward of the footpath, there is a wedge of unconsolidated sediment stacked up against the cliffs (Plate 7.9). This has been studied in detail by geologists, revealing 13 distinct layers of sediment. The lower layers, closest to the cliff, some of which are now unfortunately obscured by sea-defences, comprise angular fragments of limestone and fossils of arctic fox, elephant, hare, horse, lemming, and reindeer. These sediments indicate a generally cold environment with alternating moist and drier conditions.

Above this cold deposit, a thick sequence of dune sands occur which also contains fossils of broken seashells. This must have been deposited within a relatively warm period, because during cold episodes sea-water is locked up as ice and so sea-levels are lower. Thus, with sea-level apparently at a similar level to today, this implies the climate was also similar to today. A cooler, but wet climate is suggested by a second angular layer above the dune sand, which passes upwards into a clay layer which contains flints and artefacts of the early Bronze Age. In fact, this whole sequence has been interpreted as being deposited over the last ice age and present interglacial.

Plate 7.9 Cliffs of Carboniferous Limestone and a wedge of ice age sediments (behind the cars) on the southern flank of Brean Down.

Brean Down has been occupied by humans for thousands of years as it is well-sited as a vantage point, and with its steep cliffs is easily defended. As already mentioned, Bronze Age people lived at Brean Down and built a number of round barrows, and there is evidence that the Beaker culture existed here between 4000 and 3400 years ago. Then in the following early Iron Age a major hill fort was built and possibly inhabited by the Dobunnic tribe towards the end of the Iron Age. Brean Down did not escape the attention of the Romans and a temple, built in a Romano-Celtic style, was constructed around 340 AD. Interestingly, more than 50 years later, smaller buildings were built close to the temple, but aligned east-west, and may indicate later Christian use of the originally Roman pagan site. Later burial graves on Brean Down at around 650 AD were also aligned east-west, suggesting a prolonged religious use of the site.

Brean Down is a spectacular landform with a very interesting geological and archaeological history, where an unusual combination of upland and coastal environments occurs today, all rolled into one.

Case Study 7.1 Ammonites of Somerset

As mentioned earlier, Jurassic rocks are particularly rich in ammonites, and Jurassic rocks outcrop extensively along the coast of west Somerset, at Watchet, Kilve, and Lilstock. Here excellent, although somewhat precarious, cliff exposures exist from which ammonites can be found, although care must be taken and due attention paid to tide conditions, over-hanging cliffs, and sharp shards of hard limestone.

Between 1535 and 1545, Leland visited Keynsham and wrote of "stones figured like serpents". He was of course describing the fossils we now call ammonites. Ammonites are one of the most abundant and well-known of all fossils, and their beautiful shells, often ornamented with ribbing, have been considered aesthetic treasures for many years (Plate 7.10).

Ammonites are named after the ancient Egyptian god Ammon or Amun, who held the ram in great reverence, and so is often depicted as a rams head with twisted spiral horns which strongly resemble ammonites. Pliny the Elder, a Roman historian, was the first person to write about ammonites, and referred to them as Ammon's horns, but the first collectors of ammonites inhabited the Vogelherd Cave in southern Germany about 25 thousand years ago, where they left ammonites decorated with deep engravings.

Ammonites are an extinct class of Molluscs, being related to snails, slugs, most seashells, octopus, squid, cuttlefish, and Nautilus. It is thought that ammonites lived their entire life in the sea, swimming freely. In the geological record we only find the empty shells of ammonites preserved as fossils, but these shells were inhabited by a soft-bodied animal, which probably looked like a squid with a shell. It probably had long tentacles which protruded from the shell opening or aperture, and these tentacles probably surrounded the mouth parts and an organ that propelled the ammonite by squirting out water at high speed. Therefore, ammonites probably lived with the aperture at the bottom, and travelled through the water in a direction opposite to the way the aperture was pointing, thus swimming backwards. British geologists always illustrate ammonites in what must be an upside-down position; but this convention was adopted well before anyone knew anything about the life habits of ammonites.

Plate 7.10 Ammonites are commonly found as in the Lias (Jurassic) and often used as decoration in buildings (this one is from Evercreech).

The shell of an ammonite consists of whorls of chambers wrapped around an initial chamber known as a **protoconch**. *The shell can be smooth or ornamented with ribs and/or tubercles. Zig-zag lines may also be seen on the shell surface, known as sutures, and these show the* position of the chamber walls inside the shell. The middle part of the shell, which appears depressed, is called the umbilicus. Sometimes on the outside rim of the shell, a ridge may occur, known as a keel, and may have acted rather-like a keel does on a boat, to help guide the vessel through the water.

Ammonoids first occur in the geological record during the Devonian period, but become most abundant in British rocks during the Jurassic period. Individual species of ammonites didn't tend to last very long, and many only lasted a million years from their evolution from an ancestor species to extinction, which often involved their evolution into a descendant species. This high evolutionary turnover (a million years is a blink of an eye in geological terms) means that ammonites are extremely useful in dating rocks. This is because if one finds a fossil in a rock of a species which is known to only have existed during the earliest Jurassic, then one can date that rock to the earliest Jurassic. In this way geological time can be subdivided into ammonite zones. The last ammonites became extinct at 65 million years ago at exactly the same time as the dinosaurs and many other organisms.

Jurassic rocks occur throughout Somerset, but can be best seen exposed in cliffs from Watchet to Kilve. Also, old overgrown disused quarries and road cuttings can be found, and with a good hammer and chisel it isn't too difficult to find fine examples of ammonites. Indeed, ammonites are built into houses in some areas of Somerset.

Somerset Landscapes

Chapter 8. Where Lowland Meets the Sea.

Somerset has many lowland areas, mostly part of the extensive area known as the Somerset Levels (Figure 1.3). These lowlands are relatively new landscapes in that they were largely formed following the last ice age. The boundary between land and sea at present is quite artificial, as sea walls are commonplace and often are the only reason that large parts of the Somerset Levels aren't flooded on a daily basis. These coastlines are characterised by soft sediment landforms, such as saltmarshes and sand-dunes, and are very vulnerable to changes in sea level which, given the present concern about global warming, is a very topical issue.

8.1 Saltmarshes and mudflats of the Severn Estuary

The coast of Somerset, Gloucestershire and Gwent is characterised in many places by stretches of **mudflats** and saltmarshes. These occur within the intertidal zone, that is the area between the highest and lowest tide levels, and so are submerged and exposed regularly by the rising and falling of the tides. The Severn Estuary and Bristol Channel has a very high tidal range, up to 14.8m at Avonmouth, with the high water tides rising to 6m above Ordnance Datum (OD) in the Bristol Channel and up to 8.5m above OD in the inner Severn Estuary.

During the last ice age sea-levels were some 130m lower than today because so much water was locked up in the great continental ice sheets. As the ice age came to an end and the ice sheets began to melt, water returned to the sea and sea levels began to rise. The sea first entered the outer Bristol Channel around 8000 years ago and reached its present level some 2000 years ago. Much of the Somerset Levels at this time were characterised by extensive mudflats and saltmarshes, and it was only with the coming of the Romans and their technology that these wetlands began to be reclaimed by the building of sea defences to keep the tides out. However, seaward of these sea walls, the mudflats and saltmarshes persisted.

The mudflats and vegetated saltmarshes throughout the region possess a stepped-like appearance descending to the sea and may be subdivided into high, middle and low marsh terraces, and unvegetated mudflat (Figure 8.1). The formation of these landforms is complex and represents alternating periods of deposition and erosion. Deposition of sediment on mudflats, supplied on the flood tide, will build-up the mudflat surface, elevating it within the tidal range. As the mudflat surface is elevated there will be a threshold at which the surface is exposed by the tides sufficiently long and often enough to allow salt-tolerant plants to grow (Plate 8.1). At this point the mudflat has become a saltmarsh and if accretion of sediment continues upward and seaward, and it usually increases because of the sediment trapping ability of the plants, then the saltmarsh will mature and begin to extend seaward with a gently sloping shoreline down to the mudflats; in the Severn Estuary and Bristol Channel landward extension is prevented by the sea walls.

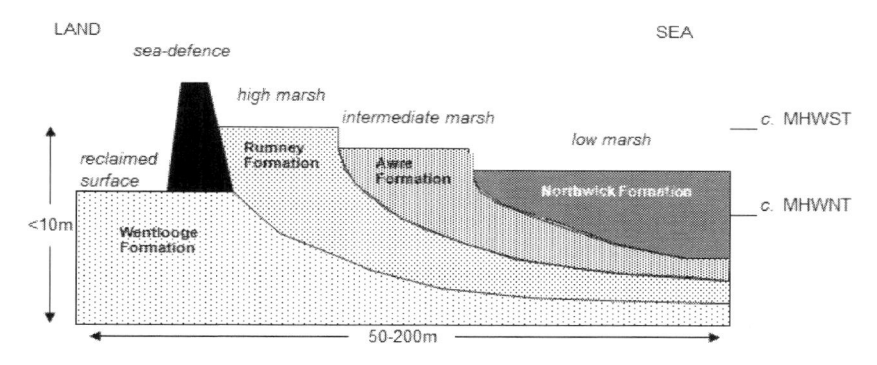

Figure 8.1 The **geomorphology** of Severn Estuary saltmarshes.

The present high marsh was formed in this way during a phase of deposition; however, its seaward extension was brought to an end by an increase in erosion along its shoreline which began to cut back into the saltmarsh, producing cliffs along the length of its shore. This erosion along the shoreline however, did not prevent sediment being deposited on the saltmarsh surface. A second phase of deposition must have ensued, banking sediment against the cliffed shoreline of the first (high) marsh, to produce the present middle marsh. A further cycle of erosion and deposition followed this to cliff the middle marsh and deposit the present low marsh. The old cliffs of the high and middle marshes are largely obscured by the build-up of the middle and low marshes respectively, but usually the tops of these cliffs can be seen and have been described as clifflets. In many places the low marsh is presently being eroded to produce cliffs (Plate 8.2), suggesting the cycle is starting again. Despite this erosion, all the present marsh surfaces continue to accrete upwards as the tide flows onto the marsh over the shoreline or through networks of tidal creeks or pills.

Plate 8.1 A saltmarsh shore in the Severn Estuary where vegetation (*Spartina anglica*) is colonising a mudflat.

The reason for these depositional-erosional cycles is not entirely clear, but it has been suggested that the periods of erosion were caused by climate changes which produced strong winds and which in turn produced higher-energy waves capable of cliffing the saltmarsh shorelines. It is interesting to note that it is generally accepted that the sediment underlying the high marsh began to accumulate no earlier than the end of the 17th century, and that the middle marsh sediments began accumulating no earlier than the late 19th century; therefore, there may be some link with the so-called Little Ice Age. The deposition of low marsh sediments is thought to be entirely 20th century.

Plate 8.2 An eroding saltmarsh shore in the Severn Estuary.

Plate 8.3 The surface of a saltmarsh in the Severn Estuary, exhibiting substantial circular salt pans.

Although mudflats and saltmarshes are low-lying (Plate 8.3) and by definition occur within the tidal range, which means they can be completely submerged and overtopped by the sea, they are an important natural coastal defence system. Sea walls are invariably built behind saltmarshes to keep out the highest tides, but often these sea walls are simply earth banks. This is because saltmarshes are excellent absorbers of wave-energy, and thus protect these simple sea walls from most of the wave action that threatens to undermine them and flood the reclaimed land behind. In areas where saltmarshes have disappeared, the sea walls tend to be specially designed and constructed of concrete with large boulders at their base, to act as saltmarsh substitutes. The main difference between these two different coastal defence strategies is that saltmarshes are free, and the engineered defences cost millions of pounds to build. It is no wonder that many authorities are now ensuring that saltmarshes are managed effectively.

Archaeologists also are now becoming interested in saltmarsh and mudflat deposits because of the excellent preservation potential of artefacts in the sediment. Below the present saltmarshes exist a number of horizons which represent ancient surfaces which people occupied. Evidence for Mesolithic cultures has been found, including human and auroch footprints, dwellings, and other artefacts. Neolithic, Iron-age, and Roman artefacts are also common, the latter particularly so, and many pieces of pottery can be found in certain places at the base of some saltmarsh cliffs where the material is being exhumed by erosional processes. Medieval and post-Medieval artefacts can also be found in the saltmarsh sediments themselves.

In addition to the geological, geomorphological, engineering and archaeological interest, the saltmarshes and mudflats of the Severn Estuary and Bristol Channel are internationally renowned havens for wildlife. Many birds visit the intertidal zone often feeding on the soft mudflats and roosting and breeding on the saltmarshes. Also, the saltmarsh plants themselves are of great interest, because for many species these marginal environments sandwiched between the land and the sea are the only places where they can survive, and they also support a great variety of invertebrate life.

In the late 1980s, and again in the first decade of the 21st century, the possibility of building a barrage across the Severn was investigated, positioned in a line between Brean Down and Llantwit Major. Once again, the government has not decided to go ahead with it yet, but if it were to be built it would have a significant effect on the mudflats and saltmarshes upstream of the barrage. First, it would reduce the tidal range, exposing less mudflat at low tide. This would reduce the feeding range of birds and force some to go elsewhere. A barrage would also reduce the amount of erosion of saltmarshes because waves would be prevented (to a degree) from entering the estuary. This might allow the further seaward extension of the saltmarshes, reducing the amount of exposed mudflat still further.

However, a bigger threat to wildlife and humans in the region is that of sea level rise associated with global warming, which is thought to be occurring here at approximately 2mm a year. Because all the mudflats and saltmarshes are backed by sea walls, they cannot migrate inland as they would under natural conditions, so with continued sea level rise they may become progressively submerged and ultimately drowned completely. Under such circumstances the sea walls would have to be built-up and strengthened at considerable cost. It is a sobering thought however, that at present most of the Somerset Levels lie below the present level of high water tides and any increase in sea level will only serve to increase the frequency and severity of the flooding that is already experienced.

Case Study 8.1 Tides in the Bristol Channel

For those who live at the coast, their relationship with the sea is strongly influenced by tides. Tides dictate when certain parts of the coast can be visited and when certain marine resources can be exploited, such as collecting shellfish. Tides also influence recreational activities at the seaside, swimming, boating, fishing, surfing, to name but a few. Also, tides can be extremely dangerous, claiming many lives around Britain each year.

Tides are in fact waves that travel very slowly around the globe. Unlike most ocean waves which are created by wind blowing over the sea-surface, tides are created through the gravitational pull from the Moon and the Sun acting upon Earth's water bodies. These astronomical bodies attract sea-water elevating the sea-surface to form what is known as a tidal bulge, and as the Earth rotates the tidal bulge remains stationary. So in fact, tidal waves do not move, but the Earth moves underneath them. (Note that the term "tidal wave" is no longer used to describe catastrophic waves, as they have nothing whatsoever to do with tides; they are now called **"tsunami"**, from the Japanese meaning "harbour wave").

A tidal bulge exists on both sides of the Earth, so most places will experience two tides a day. A rising tide is called a flood tide, whilst a falling tide is called an ebb tide. The completion of flood to ebb to flood tides (or vice versa) is considered one tidal cycle. The movement of the flood and ebb tides onto and off the coast creates tidal currents. At the start of each tidal cycle current speed is minimal, but it gradually accelerates to a peak before decelerating to achieve maximum flood or ebb. Once the maximum is reached currents stop to produce still water. This doesn't last for long and the tide and currents quickly reverse.

Tidal currents are very important for coastal landscapes, as currents have the ability to pick up, transport, and deposit sediment. Sediment is entrained and transported in the tidal flow during the accelerating phase, and then it is finally deposited during deceleration and still water. Evidence for the tidal movement of sediment is clear at many coasts, including Weston-super-Mare. Wave-like structures, such as ripples and dunes, can be seen on beaches and sand/mud flats, which are formed when tidal currents move across the sea bed (Plate 8.4).

If tides are created by the attraction of both the Moon and the Sun, then variations in the magnitude of tides is to be expected because orbits change. Indeed, there is a general cycle of higher than normal tides every two weeks, separated by a phase of lower than normal tides. These phases are called spring and neap tides respectively. Spring tides occur when the Moon and the Sun are aligned, combining their attractive forces, either with the Sun behind the Moon (as with a New Moon), or when they are on opposite sides of the Earth but still aligned (as with a Full Moon). Neap tides however, occur when the Moon and Sun are at right angles to the Earth, always occurring at the Half Moon stage.

Plate 8.4 Ripple marks in intertidal sediment at Weston-super-Mare, created by tidal currents.

The tides of the Bristol Channel and the Severn Estuary are world famous (well at least amongst geographers). This is primarily because they have the second largest tidal range in the world, only the Bay of Fundy in eastern Canada is larger. The tidal range is the vertical distance between low and high tide, and in the Bristol Channel it is approximately 14.5m. This large tidal range occurs for a number of reasons, and the funnel shape of the Severn Estuary is a major factor, constricting the tidal wave as it enters the estuary, piling the water up often forming a tidal bore. The energy created by such a tidal range has been the subject of much research recently by the government who are considering harnessing the energy through the use of tidal barrage, stretched across the estuary from Brean Down to Llantwit Major.

8.2 Stert Flats

The tidal flats of Bridgwater Bay are notoriously treacherous, comprising soft sand, silt, and mud, over which the tide comes racing in. They represent a unique environment, important for shore-birds and waders who feed here, important for their role in the sediment circulation of the Bristol Channel, and important as modern analogues for the ancient sediments underlying the nearby Somerset Levels.

Stert Flats cover the tidal area that stretches from Hinkley Point and Stolford in the southwest to the River Parrett and Burnham-on-sea to the northeast. North of the River Parrett the Berrow Flats extend up the coast to Brean Down. The main difference between the two flats is that the Berrow Flats are sandier than the Stert Flats, although both are of similar dimensions with up to 3km between the high tide and low tide marks.

Because of the international importance of Stert Flats, they have been designated a nature reserve, with excellent access to the coast. Parking is available near Walt Common and in the village of Steart, near Cox's Farm, in a designated car park associated with the nature reserve. Here information boards outline the significance of the area for wildlife and advertise the main footpaths around the area.

Steart village is located in the middle of a peninsula, with the estuary of the River Parrett behind it to the east and Stert Flats to the west and seaward. This peninsula should be considered as a spit, anchored to the coast at Stolford and running for about 6km to its tip at Stert Point and Fenning Island. Here a number of birdwatching hides have been erected. The spit tapers to the northeast as a consequence of the northeast flowing along-shore currents and River Parrett, in front and behind it respectively. It is one of only about ten major spits in the British Isles.

The spit coastline is quite a complex landform, with a number of different component parts. Stert Flats are the seaward component, characterised by unvegetated tidal flats made up of silt and mud (Plate 8.5). The surface of these flats is extensive and slopes gently up towards the shore, covering a tidal range from the low tide to the mean high water mark. Above this level, to the level of the highest tides, the flats are mostly vegetated to produce a saltmarsh component, often with a number of small yet distinct tidal channels. There is a transition between the unvegetated tidal flats and saltmarsh, with very patchy vegetation to begin with, but quickly becoming well established the higher one looks in the tidal frame.

Plate 8.5 Stert Flats in Bridgwater Bay with Hinkley Point power station in the distance. Saltmarshes and mudflats are clearly seen.

Behind the saltmarshes, out of the reach of the normal high tide levels, freshwater ponds-up to provide swampy conditions, supporting extensive swathes of reedbeds, so important for attracting and providing cover for small birds. Directly inland of these reedbeds, a raised ridge is encountered and runs the length of the seaward edge of

Somerset Landscapes

the spit. This ridge is composed of fairly small pebbles, providing a firm substrate exploited by the coastal path, which runs along the top of the ridge. The pebble ridge is fairly well vegetated and appears to be a relict feature (Plate 8.6). However, such features are usually built up through the action of storms. The normal tidal currents are not sufficiently strong enough to transport pebbles, and so it is during storm events, when wave energy is increased, that these large sediment grains may be picked up and carried further inland than would be expected under normal conditions. The source of these pebbles is not local as there are no coastal cliffs nearby, and it appears that the pebbles originated further west along the Somerset coast, from the shale and limestone cliffs at Lilstock, and from the supply of Quantock sandstone to the coast by rivers and streams. These would have been moved eastward along the coast by waves and currents for inclusion into Stert Spit during storms. The fact that the pebble ridge now looks relict, and also the establishment of reedbeds in front of it, suggests that the storms that created the ridge did not occur recently.

Plate 8.6 The largely relict shingle barrier on Stert Spit.

The pebble ridge represents the limit of the modern coast, for behind it the land is turned over to agriculture and farms. However, within a number of fields around the village of Steart, distinct ridges can be seen, mostly orientated parallel to the coastline, and may be either very old pebble ridges or sand dunes (although it is difficult to look at the sediment that make up the ridges due to the complete lack of exposures). The presence of the fossil ridges suggest that either storms are much less severe than they used to be, or that the coastline is advancing seaward, leaving these ridges stranded high and dry as it migrates. This kind of coastal **progradation** is common in areas where sea-level is relatively stable and there is an abundant sediment supply, both of which apply to this area.

Further inland still, the landscape becomes almost completely flat and large extensive fields of cereal crops are found (Plate 8.7). These flatlands are part of the Somerset Levels, which are essentially prehistoric mudflats and saltmarshes that were either reclaimed by the Romans or dried out when natural coastal barriers, such as the pebble ridges, formed preventing the sea from inundating the area. The sediment directly underlying the Levels here is tidal silts and clays, very similar to the sediment being deposited on Stert Flats today, only some 2000 to 3000 years older.

Plate 8.7 Flat agricultural landscape near Steart, part of the low-lying Somerset Levels.

8.3 Somerset Levels and sea-level change

Throughout the last ice age, sea-levels around the British Isles were some 130m below the current level, and Somerset was well and truly landlocked. The sea lay many miles to the west, somewhere on a line between southern Ireland and Cornwall.

During the last ice age northern Britain was heavily depressed under the weight of ice sheets and glaciers and, like a see-saw, southern Britain was uplifted as a consequence. As the ice sheets began to melt towards the end of the last ice age, northern Britain started to rise back up as the weight of ice was removed, and according to the see-saw analogy, southern Britain began to subside.

Also, with the melting of the ice, the water that was locked up in ice sheets and glaciers returned to the sea, thus causing sea-levels to rise. Furthermore, the climatic warming that occurred at the end of the ice age raised the temperature of the ocean, thermally expanding sea-water so further contributing to sea-level rise. With sea-levels rising and southern Britain sinking, it was only a matter of time before the sea reached its current level.

The sea first entered the Bristol Channel around 8,000 years ago and quickly flooded the western part of Somerset, to produce the area we now call the Somerset Levels. This area, once a flooded inlet of the sea, is characterised by a flat, level landscape. This is because the type of coastal environment that was created was that of mudflats and saltmarsh, very similar to the present day Bridgwater Bay, Severn Estuary and Bristol Channel. If one looks at the sides of newly cut drainage ditches around the Levels, it is possible to see the sediment that makes up the flats. It is typically a blue-grey coloured clay and often contains fossils of marine animals, such as seashells. This blue clay is found throughout the Somerset Levels.

In some places, such as in the Brue Valley, peat can be found at the surface, sandwiched between the blue clay to the west and hills to the north, south and east. This peat was formed because freshwater running off the hills was prevented from reaching the sea by the build up of blue clay, and so the freshwater ponded up to produce conditions wet enough for peat growth. It is on these peat surfaces that Neolithic people built their trackways, such as the famous Sweet Track (Plate 8.8), to gain access to the peat bogs and communicate between island settlements At certain times after the ice age, the advancing sea paused or retreated slightly before continuing to rise, and on these occasions the peat extended westwards towards the sea. We know this because layers of peat often occur within the blue clay.

Plate 8.8 A replica of the Sweet Track at Shapwick Heath in the Brue Valley.

Today, the landscape underlain by the blue clay, lies around 6m above present day mean sea-level, and the area underlain by the peat is around 5m. This shows that the peat did accumulate in a low area behind the blue clay, which prevented the freshwater from reaching the sea. This also has a more alarming side to it. The present day tidal range in the Severn Estuary and Bristol Channel extends in some places up to 9m above mean sea-level, that is well above the ground surface elevation of the Somerset Levels. Indeed, the area would still be flooded twice daily at high tide, but for the sea-walls and river levées that were first erected by the Romans to keep out the sea and reclaim the land for agriculture (Plate 8.9).

Plate 8.9 The narrow Somerset coastline near Berrow separating the sea from the Somerset Levels.

This is a cause for concern because of the threat of global warming from an enhanced greenhouse effect. Global temperatures are rising and sea-levels are rising as a consequence. This is mainly because the higher temperatures are melting mountain glaciers, in places like the Alps, and the oceans are warming up, and as they warm up they expand. There is currently little evidence to suggest the polar ice caps are melting to any great extent, indeed it appears that Antarctica is growing in size because more snow is falling there.

Global sea-levels are rising, but southern Britain in general is at a greater risk of coastal flooding than northern Britain. This is because southern Britain is still sinking as an after effect of the ice age, although this may only be slight in Somerset. Furthermore, with global warming we can expect more storms and higher rainfall in the winter. Somerset rivers have substantial catchments, draining large areas; thus, when high rainfall is accompanied by storms and increasingly higher tides, there is nowhere for the water to go, except over the sea-walls (Plate 8.10) and levées to flood the Levels which were once part of the natural coastal environment. It is likely that, as global warming proceeds, flooding events will become more frequent and more damaging.

Plate 8.10 Sea-defences (a wave-return wall) at Brean Down, part of Somerset's coastal defences.

The impact of sea-level rise on the Somerset coastline largely depends on the type of coastline present. For example, saltmarshes and sand dunes may become cliffed and/or submerged by rising sea-levels and increased storminess, whereas gravel shorelines become breached and migrate inland. There are a number of strategies which could be adopted to deal with the problems of coastal erosion and flooding. In valuable developed areas, whether residential or industrial, it may be most appropriate to protect the coastline by constructing hard structures, such as sea-walls, embankments, groynes, or tidal barriers. However, this is extremely expensive, so in areas where communities are already accustomed to flooding, it has been suggested that people should adapt by elevating buildings, modifying existing drainage systems, and improving storm and flood warnings. Both these solutions are short term measures (typically only effective up to 100 years) and interrupt the natural system, so wherever possible (mainly on undeveloped land, or areas with low housing density) the coastline should be allowed to retreat naturally (Plate 8.11). This would be very unpopular and government would have to play a strong role in buying up threatened land and prohibiting future development.

Plate 8.11 Temporary setback of agricultural land at Porlock in response to the landward migration of the storm beach.

Whatever strategy is adopted it is now almost certain that global warming is occurring, perhaps because of enhancement of the greenhouse effect, and that as a consequence sea-levels are rising and storm frequency and intensity is also increasing. We are committed to sea-level rise into the 21st century, because even if we banned all known greenhouse gas emissions now, it would be another 50 years or so before the ocean responded to stabilise sea-levels. If we carry on emitting greenhouse gases at the present rate the Somerset Levels may ultimately be reclaimed by the sea.

Case Study 8.2 Detecting sea-level change

In 1998, myself and colleagues in the Department of Geography at Bath Spa University published the results of a 3 year study of sea-level change in the Somerset Levels. The results were published in the scientific journal The Holocene, an international journal focusing on recent environmental change. The study was part of an international project set up by UNESCO to investigate coastal change on a global scale.

The construction of curves depicting sea-level change since the last ice age requires the identification of Sea-Level Index Points (SLIPs). These are points for which altitude, age, indicative meaning and range, and sea-level tendency are known (these terms are explained below). In the United Kingdom altitude is usually given in relation to Ordnance Datum at Newlyn (OD). Age may be determined using the radiocarbon dating method if sufficient organic matter (eg wood, peat, shells) is available. Indicative meaning refers to the tidal level that a SLIP represents (eg Mean High Water Spring Tides (MHWST) or Highest Astronomical Tides (HAT), etc), and indicative range encompasses any uncertainty present in assigning an indicative meaning (eg MHWST-HAT). Sea-level tendency indicates the sea-level trend at the SLIP, whether sea-level is falling or rising, and is expressed as negative or positive tendency respectively.

A number of sea-level curves have been constructed using alternating marine clays and freshwater peats in the Somerset Levels (Figure 8.2). However, each of these models has been constructed using age and altitude information only, with no published details of indicative meaning or tendency of their age/altitude points, and thus they cannot be considered as SLIPs. The aim of our study was to make an evaluation of these models, based on a study at Nyland Hill in the Axe Valley, near Cheddar (Plate 8.12). Our study suggests that the accuracy of the earlier models is questionable.

Somerset Landscapes

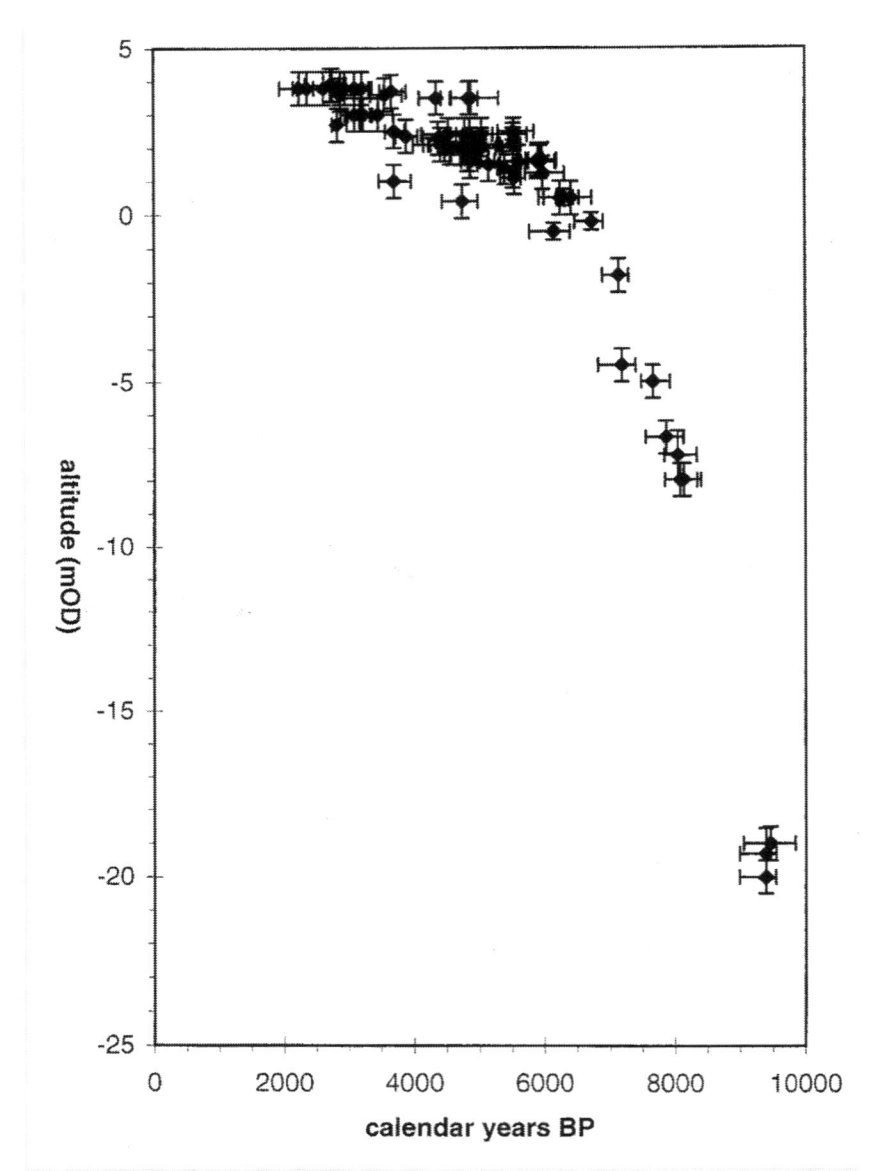

Figure 8.2 A record of post-glacial sea level rise in the Somerset Levels.

Plate 8.12 Surveying the **Holocene** sediment of the Somerset Levels requires a number of instruments. Here, my colleague Dr Rick Curr (Bath Spa University) is using a seismograph at Nyland Hill to survey the shape of the bedrock under the Holocene sediment.

The main findings of our study stem from the fact that we can now demonstrate that sediment can become severely compacted in the Somerset Levels, so lowering the original altitude and giving a false impression of where sea-level was at a particular time. Because the severity of compaction has hitherto been unrecognised it has led to a number of misconceptions.

First, that following the ice age sea-level rose very rapidly to flood the Somerset Levels; secondly, in order to accommodate such a rapid rise it has been suggested that the region is generally subsiding; and thirdly, that sea-level ceased rising about 4000 years ago and attained roughly its present level. We now suggest that the very rapid rise in sea-level and flattening out of the sea-level curve at 4000 years ago are both artefacts of SLIP altitudes lowered by compaction, and that when one restores the correct altitude there is no evidence for any significant subsidence in the region.

This is important because of the threat of global warming from an enhanced greenhouse effect. Global temperatures are rising and sea-levels are rising as a consequence, but Somerset was considered to be at a greater risk of coastal flooding than other areas of Britain because it was thought to be still subsiding as an after effect of the ice age. However, it now appears from our study that that is one factor we can remove from the equation as far as Somerset is concerned. Although, we still have to worry about sea-level rise, increasing storminess with its rain and low atmospheric pressure, all of which will increase the frequency and magnitude of coastal flooding in the Somerset Levels.

Case Study 8.3 The 1607 flood – a tsunami in the Bristol Channel?

Tsunami don't occur in Britain – do they? Well, yes they do, and one can strike any coast at any time, but they are very rare events indeed.

Tsunami are large waves caused by either submarine earthquakes, landslides, volcanic eruptions or by a comet plunging into the Ocean. They are common in the Pacific Ocean because of the major fault lines and volcanoes that occur there, and as a consequence a Tsunami Warning System has been created across the Pacific. They are less common in the Indian Ocean, and rarer still in the Atlantic.

The Asian tsunami that occurred on 26th December 2004 has shown that although tsunami may not be common, when they do strike they can be devastating and, in hindsight, the benefits of a warning system are painfully obvious. The same is true of the North Atlantic where tsunami are rare, but no less severe.

In 1755, a magnitude 9.0 earthquake (almost the same as the Asian earthquake) occurred in the seabed offshore Portugal that sent a 15 m (49ft) high tsunami that struck Lisbon probably killing close to 50,000 people. The tsunami radiated outwards to damage coastal areas in Spain, Morocco and the Caribbean islands some 5,700 km away – a truly pan-North Atlantic event. It also affected southwest Britain, being recorded as far north as Swansea.

An earlier catastrophic coastal flooding event occurred in Somerset, and neighbouring north Devon, Gloucestershire and south Wales, in 1607 that appears to have killed around 2000 people and plunged the region into social and economic chaos. Historians have usually attributed the flooding to a storm but new geographical research, that myself and Dr Ted Bryant of the University of Wollongong in Australia have undertaken since 2002, has shown that it was more likely to have been a tsunami. We published our initial ideas in 2002 in the scientific journal Archaeology in the Severn Estuary.

According to contemporary accounts, the flood occurred rapidly in apparently good weather on 20th January 1607 "for about nine of the morning, the same being fayrely and brightly spred, many of the inhabitants of these countreys prepared themselves to their affayres" when they saw "mighty hilles of water tombling over one another in such sort as if the greatest mountains in the world had overwhelmed the lowe villages" – a description closer to a tsunami than a storm.

In summer 2004, Ted and I travelled the length of the Bristol Channel coast from Barnstaple to Gloucestershire to the Gower looking for signs that a tsunami had struck the coast. We were filmed doing this for a BBC2 Timewatch programme 'Killer Wave of 1607' originally scheduled for broadcasting in February 2005, but following the Asian disaster its showing was postponed until the Spring to allow some re-editing due to the sensitive nature of the topic.

Somerset Landscapes

Figure 8.3 A woodcut depicting the impact of 1607 flood from one of the contemporary accounts.

From our fieldwork we estimated that a 1607 tsunami would have been at least 5.5m (18ft) high when it struck the Somerset coast, travelling at a speed of 32mph, and penetrating inland as a moving wave for at least 2½ miles from the shore. These results are consistent with the description of "mighty hilles of water" and others that the wave is "affirmed to have runne .... with a swiftness so incredible, as that no gray-hounde could have escaped by running before them". However, the historical accounts tell us that the waters reached the foot of Glastonbury Tor some 14 miles inland where it flooded alms houses - this is possible as the ground in the Somerset Levels slopes landward, so that when the tsunami wave broke and collapsed the water rushed further inland rather than return to the sea.

The accounts contain a long list of Somerset places that were badly affected, including Berrow, Yatton, Puxton, Congresbury, Kingston Seymour, Worle, Kewstoke, Banwell, Wick, Weston-Super-Mare, Uphill, Kenn, Combwich, Burnham, Lympsham, East Brent, Mark and Brean, the latter four are said to have been "swallowed up" by the wave with 26 fatalities recorded at Brean.

Amongst the Brean victims, a John Good lost his wife, five children, and nine servants as the wave struck, but he saved himself by clinging to thatch that carried him for more than a mile before it washed up on a bank. Thatch seems also to have saved the wife and son of a John Stowe of Berrow, upon which they were washed two miles by the wave to safety; he was not so lucky and drowned with three of his other children. At Bridgwater, "two villages near theirabouts and our market town overflown [by the wave] and report of 500 persons drowned, besides many sheep and other cattle".

At Kingston Seymour a plaque in the church commemorates the event: "an inundation of the Sea-water by overflowing and breaking down the Sea banks; happened in this Parish of Kingstone-Seamore, and many other adjoining; by reason whereof many Persons were drown'd and much Cattle and Goods, were lost: the water in the Church was five feet high and the greatest part lay on the ground about ten days". The water lay on the ground for such a long time partly because of the landward slope of the ground, and probably because all the sluice-gate keepers had perished.

Apart from the historical accounts and our field evidence, the potential smoking gun that would confirm in my mind that the 1607 flood was caused by a tsunami is a report that a previously overlooked account states that an earth tremor was felt on the morning in question. It is possible that an active fault system offshore southern Ireland had a significant earthquake that created the tsunami. Indeed, the fault in question has apparently experienced a magnitude 4.5 earthquake in 1980, not big enough to cause a tsunami, but indicates that it may have been able to produce a bigger one in the past. Part of our ongoing research is to establish the frequency of earthquakes and tsunami that have struck the Somerset coast through time, but we know that 1607 was seismically active locally with definite accounts of earthquakes being felt in the Bristol Channel area in February and May of that year.

What has happened before could happen again, and the low-lying areas are now more populated than in 1607, so casualties could be higher. But for a tsunami to have any impact on the Somerset coast the tsunami must arrive close to high tide, because the tide goes out so far in the Bristol Channel, if the tsunami arrived at low tide it might not reach the shore, let alone flood the land. The 1607 event shows that these unfortunate coincidences do occur and for this reason the Bristol

Channel coast, as well as all coasts everywhere, should be protected by a tsunami warning system.

Without such a warning system, people living in low-lying coastal areas should go upstairs, or climb a tree, if they feel an earth tremor or see the sea rise or rush out suddenly, as research has shown that many stone buildings and trees withstand all but the biggest tsunami. Stay up off the ground for at least a few hours as it is likely that more than one tsunami wave will strike, the second or third wave often being higher than the first.

Glossary

ammonite - an extinct marine cephalopod mollusc closely related to modern squid

***Anabacia* Limestone** - a limestone rich in *Anabacia* coral, the youngest bed of Inferior Oolite (Middle Jurassic) found at Doulting

andesite - a grey volcanic rock characterised by the presence of the minerals oligoclase and/or andesine, and moderate amounts of silica; the lava is therefore, of moderate viscosity capable of flowing some distance, but usually as discrete flows

anticline - a fold in strata that is convex-up, created by compression usually associated with orogenic activity

asthenosphere - the upper part of the earth's mantle

Avill Group - the lowest subdivision of the Ilfracombe Beds, comprising slates, siltsones and sandstones and the Cockercombe Tuff, from the Devonian of the Quantocks

basalt - a dark volcanic rock, low in silica, often forming the ocean-floor (SiMa) and on oceanic islands; the lava is of low viscosity and flows very easily with different flows often merging together

belemnite - an extinct marine cephalopod mollusc shaped like a bullet

Belemnite Marls - pale to dark grey mudstones with abundant belemnites, from the Lower Jurassic (Lower Lias) of Dorset

biomicrite - fine-grained limestone derived from biological material eg shells

biostratigraphy - the subdivision of geological time based on the occurrence and succession of fossils in rocks

Black Ven Marls - dark grey mudstones from the Lower Jurassic (Lower Lias) of Dorset

Blue Lias - formally known as the Blue Lias Formation, it comprises alternating beds of limestone and shale of the Lower Jurassic (Lias)

brachiopod - a marine mollusc that is commonly found as a fossil, but living brachiopods are now rare.

breccia - a rock-type characterised by the inclusion of angular fragments of other rocks

Bridport Sands - previously restricted to Dorset, it is now formally incorporated into the more extensive Bridport Sand Formation (which now includes the Cotteswold Sands, Midford Sands, Yeovil Sands, and Cotswold Sands as well), it is a yellow-brown silty sandstone from the Lower Jurassic (Lias)

Burtle Beds - marine and terrestrial interglacial Pleistocene deposits found in the Somerset Levels

Carboniferous Limestones - includes variously named hard fossiliferous limestones, formally known as the Carboniferous Limestone Supergroup

chert - bands of silica found in sedimentary rocks

Cockercombe Tuff - a green-coloured deposit derived from volcanic activity, a constituent of the Avill Group from the Devonian of the Quantocks

coccoliths - microscopic plate-like fossils of planktonic algae

colluvium - any accumulation of sediment transported by gravity, ie downslope movement

compressional stress - the stress suffered when two crustal plates collide, often resulting in folding, orogeny and reverse (thrust) faulting

crinoid - an echinoid that is commonly referred to as a sea-lily, often found as fossils, especially disarticulated pieces of the stem

cuesta - a ridge comprising a gentle dip-slope and a steep scarp-slope (synonymous with escarpment)

Cutcombe Slate - grey to brown slates with a band of limestone (Rodhuish Limestone) from the Devonian of the Quantocks

diagenesis - refers to processes that operate on sediment following deposition, such as sediment compaction and cementation, leading to the sediment becoming a rock

Dinantian Series - a series of rocks from the Lower Carboniferous, includes the Carboniferous Limestone

Ditcheat Clay - Lower Jurassic (Lower Lias) silt and clay 40-60m thick, found in the Evercreech area

Dolomitic Conglomerate - a terrestrial deposit (really a breccia containing angular rock fragments) derived from the erosion of upland, following the Variscan Orogeny, during the Permo-Trias

Doulting Conglomerate - a thin (50cm thick) limestone containing pebbles, indicating a fossil shoreline, from the Middle Jurassic (Inferior Oolite) of Doulting

Doulting Ragstone - an oolitic limestone containing abundant fossils, from the Middle Jurassic (Inferior Oolite) of Doulting

eclogite - a metamorphic rock formed under high temperatures and pressure, often associated with the upper mantle (asthenosphere)

endogenetic processes - earth processes that are driven by energy from within the earth, ie geothermal heat

exogenetic processes - earth surface processes that are driven by energy originating from beyond the earth, ie solar radiation and the gravitational effects of the sun and moon (eg tides)

extensional stress - the stress suffered when two crustal plates diverge, often resulting in normal faulting and rifting

footwall - rock that occurs below a fault plane

foraminifera - a microscopic single-celled animal that secretes a calcareous or organic-based shell which is commonly preserved in sedimentary rocks

gabbro - the coarse-grained equivalent rock-type to basalt, often occurring at the base of the SiMa

galena - lead ore

Gault - pale to dark grey mudstone, contains fossiliferous nodules, from the Cretaceous Period

geomorphology - the study of the form of the earth's surface

glacial - a cold period in time when glacial ice advances to increase its extent, usually as ice sheets; and also refers to the action of ice on the landscape

glauconite - a green mineral, related to mica, found in marine sedimentary rocks

granite - a coarse-grained igneous rock typical of continental crust (SiAl)

Green Ammonite Beds - grey mudstone with ammonite bearing limestone bands, from the Lower Jurassic (Lower Lias) of Dorset

Greensand - soft sandstone, green colour produced by the presence of the mineral glauconite, contains chert, from the Cretaceous Period

hanging wall - rock that occurs above a fault plane

Hangman Grits - includes a number of subdivisions representing various rock-types (slates to conglomerates) deposited in a shallow water environment, from the Devonian Period

Harptree Beds - a Jurassic deposit on the Mendips that has been silicified by hydrothermal mineralisation

head - sediment containing angular rock fragments, derived from solifluction processes in a periglacial environment

Hettangian - the earliest epoch of the Jurassic

Hodders Combe Beds - the uppermost subdivision of the Hangman Grits, consisting of approximately 300m of sandstones and conglomerate

Holocene - the most recent geological epoch spanning 10,000 years ago to the present day (synonymous with post-glacial)

hydrothermal mineralisation - the deposition of mineral ores by superheated water

Ilfracombe Beds - a subdivided group of rocks consisting of slates and limestones of the Devonian Period

inlier - an area of older rocks surrounded by younger rocks, often in the core of an anticline

interglacial - a warm stage during the Quaternary Period (ice ages) when temperatures were at least as warm as today

intertidal zone - the area between the low and high tide levels

Ipswichian Interglacial - the last interglacial stage recognised in the British Isles, dated to around 125,000 years ago; the climate was warmer than today

karst - features formed by the dissolution of limestone eg caves, swallow-holes, etc

Leighland Beds - the uppermost subdivision of the Devonian Ilfracombe Beds, mainly slates and sandstones, but with some limestone

Lias - formally known as the Lias Group, spanning the latest Triassic to Lower Jurassic, and includes the Blue Lias

lithification - the process whereby unconsolidated sediment becomes rock

Mercia Mudstone Group - red mudstones and siltsones, part of the New Red Sandstone of the Triassic Period

Midford Sands - see Bridport Sands

Milankovitch Cycles - regular astronomical cycles that are known to influence climate on earth; includes the 100,000 year eccentricity cycle, 40,000 year obliquity (tilt) cycle, and the 21,000 precessional cycle

Moho Discontinuity - the boundary between the earth's mantle and crust

Morte Slates - formally known as the Morte Slate Formation, it is a sequence of maroon slates approximately 1200m thick, from the Devonian

mudflat - an area of unvegetated mud in the intertidal zone

mudstones - very fine-grained rock formed through the lithification of mud

Old Red Sandstone - deltaic sandstones from the Devonian, includes the Portishead Beds

oolite - a limestone comprised of ooids, which are small pellets which concentrically grow by being washed around in warm shallow lime-rich seas

orogeny - the process of mountain building caused by the convergence of crustal plates

outlier - an area of young rock surrounded by older rock, often in the core of a

Pennard Sands -formally known as the Pennard Sand Member, it is a Lower Jurassic (Middle Lias) sandy clay from the Evercreech area

pericline - a dome-like anticlinal fold

peridotite - a high density rock rich in olivine that forms in the mantle

Permo-Trias - a combined reference to the Permian and Triassic Periods which has arisen in the past because of the difficulty in Britain of assigning rocks to either period

Pleistocene - the first epoch of the Quaternary Period, spanning 1.81 million to 10,000 years ago

Portishead Beds - sandstones and conglomerates of the Old Red Sandstone (Devonian)

progradation - the seaward advance of a coastline as the result of high sediment accumulation

protoconch - the earliest-formed part of an ammonite or belemnite (usually small and bulbous)

Pylle Clay - formally known as the Pylle Clay Member, it is a Lower Jurassic (Lower Lias) dark grey mudstone from the Evercreech area

Quaternary - the most recent geological period spanning 1.81 million years ago to the present day

radiolaria - a microscopic single-celled animal that secretes a siliceous shell which can be preserved in sedimentary rocks

Roadwater Limestone - a bed of limestone within the Ilfracombe Beds, occurs between the Cutcombe Slate and Leighland Beds, from the Devonian of the Quantocks

Rodhuish Limestone - a limestone occurring within the Cutcombe Slate, from the Devonian of the Quantocks

Rodway Beds - siltstones and sandstones from the Lower Carboniferous (Namurian) of the Quantocks, formally known as the Rodway Siltstones Formation

rhyolite - a fine-grained volcanic rock that is rich in silica and is therefore, highly viscous, not flowing easily as a lava

saltmarsh - an area high in the intertidal zone colonised by salt-tolerant plants that help to trap sediment

scree - also known as talus, this is the accumulation of angular rock fragments derived from the mechanical weathering of rocks, commonly seen at the foot of cliffs

shale - a fine-grained sedimentary rock similar to mudstone, but with fine laminations

Shales with Beef - dark grey mudstones with bands of fibrous limestone known as beef, from the Lower Jurassic (Lower Lias) of Dorset

SiAl - abbreviation for continental crust, as continental rocks are rich in silica (Si) and aluminium (Al)

silicified - fossils that have been replaced by silica, or sediment that has been cemented by silica. such as siliceous soils known as silcrete

siltstones - sedimentary rock made up of silt-sized particles

SiMa - abbreviation for oceanic crust, as continental rocks are rich in silica (Si) and magnesium (Ma)

Sinemurian - an epoch of the Lower Jurassic

solifluction - the downhill transport of soil, often a very slow process

Spargrove Limestone - muddy limestone from the Lower Jurassic (Lower Lias) of the Evercreech area

sphalerite - zinc ore

sponge spicules - living sponges possess a delicate skeleton, made up of needle-like spicules, which falls apart after death

subduction zone - the area at the edge of a crustal plate where another plate converges/collides and descends, either downwards into the mantle or to rest under the more buoyant plate; associated with orogenic activity

swallet - synonym of swallow-hole

swallow-hole - a hole in a limestone landsurface produced by dissolution and/or the collapse of an underground cave; rivers/stream often descend into swallow-holes and flow underground (a karst feature also known as a sinkhole, swallet or doline)

swash - the up-beach movement of water as a wave breaks

syncline - a fold in strata that is concave-up, created by compression usually associated with orogenic activity

tectonics - the study of the earth's structure and processes relating to this structure, such as continental drift and plate tectonics, that can result in volcanic activity, and the folding and faulting of rocks

terracettes - small stair-like features on hillslopes, formed as the result of downhill soil movement

trilobite - an extinct arthropod from the Palaeozoic

Triscombe Beds - the middle subdivision of the Hangman Grits, comprising green sandstones and mudstones, from the Devonian of the Quantocks

tsunami – a wave created by a displacement of the water column eg by an earthquake or submarine slide. The resulting wave can be large and pose a hazard to coastal communities

tufa – a calcium carbonate precipitate like 'limescale' that occurs where limestone is dissolved and then redeposited out of solution

unconformity - a period of time for which no sediment has been deposited, eg in the east Mendips, Carboniferous rocks are unconformably overlain by Jurassic rocks, and so the intervening Permo-Trias time period is represented

Variscan Orogeny - the period of mountain building that affected northwest Europe and eastern America at the end of the Carboniferous Period; the compressional stress associated with this orogeny is responsible for many of the folds and faults seen in Somerset. This orogenic event is also often referred to as the Hercynian Orogeny or the Armorican Orogeny by some authors

wadi - a valley in a semi-arid region that only occasionally has water flowing in it, often as flash-floods

White Lias - a porcellaneous white limestone from the late Triassic

Wolstonian Glaciation - the penultimate glacial stage of the Quaternary Period, to which glacial activity in Somerset has been attributed.

Bibliography

Allen, J. R. L., 1993. Muddy alluvial coasts of Britain: field criteria for shoreline position and movement in the recent past. *Proceedings of the Geologists' Association*, **104**, 241-262.

Allen, J. R. L., & Haslett, S. K., 2002. Buried salt-marsh edges and tidal-level cycles in the mid Holocene of the Caldicot Level (Gwent), South Wales. *The Holocene*, **12**, 303-324.

Allen, J. R. L., & Haslett, S. K., 2006. Granulometric characterization and evaluation of annually banded mid-Holocene estuarine silts, Welsh Severn Estuary (UK): coastal change, sea level and climate. *Quaternary Science Reviews*, **25**, 1418-1446.

Allen, J. R. L., & Haslett, S. K., 2007. The Holocene estuarine sequence at Redwick, Welsh Severn Estuary Levels, UK: the character and role of silts. *Proceedings of the Geologists' Association*, **118**, 157-185.

Allen, J. R. L., & Haslett, S. K., 2007. A wooden fishtrap in the Severn Estuary at Northwick Oaze, South Gloucestershire. *Archaeology in the Severn Estuary*, **17** (for 2006), 169-173.

Allen, J. R. L., Haslett, S. K., & Rinkel, B. E., 2006. Holocene tidal palaeochannels, Severn Estuary Levels, UK: a search for granulometric and foraminiferal criteria. *Proceedings of the Geologists' Association*, **117**, 329-344.

ApSimon, A. M., Donovan, D. T., & Taylor, H., 1961. The stratigraphy and archaeology of the Late-Glacial and Post-Glacial deposits at Brean Down, Somerset. *Proceedings of the University of Bristol Spelaeological Society*, **9**, 67-136.

Aston, M. A., & Burrow, I. C. G. (eds.), 1982. *The Archaeology of Somerset: a review to 1500AD*. Somerset County Council, 146pp.

Baily, W. H., 1864. On the occurrence of fish-remains in the Old Red Sandstone of Portishead, near Bristol. *Geological Magazine*, **1**, 293.

Baker, A., & Simms, M. J., 1998. Active deposition of calcareous tufa in Wessex, UK, and its implications for the 'late-Holocene' tufa decline. *The Holocene*, **8**, 359-365.

Barrington, N., & Stanton, W. I., 1972. *The complete caves of Mendip* (2nd Ed.). Cheddar Valley Press, Cheddar, 155pp.

Bell, M., 1990. *Brean Down excavations 1983-1987*. English Heritage, London, 278pp.

Bell, M., Allen, J. R. L., Buckley, S., Dark, P., & Haslett, S. K., 2003. Mesolithic to neolithic coastal environmental change: excavations at

Goldcliff East, 2002. *Archaeology in the Severn Estuary*, **13** (for 2002), 1-29.

Bessa, J. L., & Hesselbo, S. P., 1997. Gamma-ray character and correlation of the Lower Lias, SW Britain. *Proceedings of the Geologists' Association*, **108**, 113-129.

Biron, E., 1999. Tufa springs in Somerset. *Nature in Somerset*, 32-36.

Boulton, W. S., 1904. On the igneous rocks at Spring Cove, near Weston-Super-Mare. *Quarterly Journal of the Geological Society, London*, **60**, 158-168.

Bradshaw, R., 1966. The Avon Gorge. *Proceedings of the Bristol Naturalists Society*, **31**, 203-220.

Briden, J. C., & Daniels, B. A., 1999. Palaeomagnetic correlation of the Upper Triassic of Somerset, England, with continental Europe and eastern North America. *Journal of the Geological Society, London*, **156**, 317-326.

Bristow, C. R., & Westhead, R. K., 1993. Geology of the Evercreech-Batcombe district (Somerset). 1:10,000 sheets ST 63 NW and ST 63 NE. *British Geological Survey Technical Report WA/93/89*, 32pp.

Bristow, C. R. and others, 1999. *The Wincanton district - a concise account of the geology*. Memoir of the British Geological Survey, No. 297.

Bryant, E. A., & Haslett, S. K., 2003. Was the AD 1607 coastal flooding event in the Severn Estuary and Bristol Channel (UK) due to a tsunami? *Archaeology in the Severn Estuary*, **13** (for 2002), 163-167.

Bryant, E. A., & Haslett, S. K., 2007. Catastrophic wave erosion, Bristol Channel, United Kingdom: impact of tsunami? *Journal of Geology*, **115**, 253-269.

Buchanan, R. A., & Cossons, N., 1969. *Industrial Archaeology of the Bristol Region*. David & Charles, Newton Abbot, 335pp.

Butler, M., Williams, B. P. J., & Bradshaw, R., 1972. A new exposure of the Old Red Sandstone - Lower Limestone Shale transition at Portishead, Somerset. *Proceedings of the Bristol Naturalists Society*, **32**, 151-155.

Colborne, G. J., Gilbertson, D. D., & Hawkins, A. B., 1974. Temporary drift exposures on the Failand Ridge. *Proceedings of the Bristol Naturalists Society*, **33**, 91-97.

Coleman, A. M., & Balchin, W. G. V., 1959. The origin and development of surface depressions in the Mendip Hills. *Proceedings of the Geologists' Association*, **70**, 291-309.

Copp, C. J. T., Taylor, M. A., & Thackray, J. C., 1997. Charles Moore (1814-1881), Somerset geologist. *Proceedings of the Somerset Archaeological and Natural History Society*, **140**, 1-36.

Cox, B. M., Sumbler M. G., & Ivimey-Cook, H. C., 1999. A formational framework for the Lower Jurassic of England and Wales (onshore area). *British Geological Survey Research Report*, RR/99/01.

Crabtree, K., & Maltby, E., 1975. Soil and land use change on Exmoor: significance of a buried profile on Exmoor. *Proceedings of the Somerset Archaeology and Natural History Society*, **119**, 38-43.

Curran, P. J., 1979. The form of the peat-blue clay boundary on the Somerset Levels. *Proceedings of the Somerset Archaeology and Natural History Society*, **123**, 1-3.

Darbyshire, E. J., & West, J. R., 1993. Turbulence and cohesive sediment transport in the Parrett Estuary. In: Clifford, N.J. *et al.* (eds.) *Turbulence: perspectives on flow and sediment transfer.* (Wiley), pp 215-247.

Davies, D. K., 1969. Shelf sedimentation: an example from the Jurassic of Britain. *Journal of Sedimentary Petrology*, **39**, 1344-1370.

Davies, P., Haslett, S. K., Davies, C. F. C. and Thomas, L., 2001. Reconstructing Holocene landscape change on Mendip: the potential of tufa deposits. *Proceedings of the Cotteswold Naturalists' Field Club*, **42** (1), 42-47.

Davies, P., Haslett, S. K., Lewis, J. & Reeves, S., 2006. Tufa deposits and archaeology in the Mendip area, Somerset. In: C. O. Hunt & S. K. Haslett (eds) *The Quaternary of Somerset: field guide.* Quaternary Research Association, Cambridge, 57-66 (refs 205-236).

Donovan, D. T., 1956. The zonal stratigraphy of the Blue Lias around Keynsham, Somerset. *Proceedings of the Geologists' Association*, **66**, 182-212.

Donovan, D. T., 1958a. Easter Field Meeting: the Lower and Middle Jurassic rocks of the Bristol District. *Proceedings of the Geologists' Association*, **69**, 130-140.

Donovan, D. T., 1958b. The Lower Lias section at Cannards Grave, Shepton Mallet, Somerset. *Proceedings of the Bristol Naturalists Society*, **29**, 393-398.

Donovan, D. T., 1969. Geomorphology and hydrology of the central Mendips. *Proceedings of the University of Bristol Spelaeological Society*, **12**, 63-74.

Donovan, D. T., Bennett, R., Bristow, C. R., Carpenter, S. C., Green, G. W., Hawkes, C. J., Prudden, H. C., & Stanton, W. I., 1989. Geology of a gas pipeline from Ilchester (Somerset) to Pucklechurch (Avon), 1985. *Proceedings of the Somerset Archaeological and Natural History Society*, **132** (for 1988), 297-317.

Down, C. G., & Warrington, A. J., 1971. *The History of the Somerset Coalfield*. David & Charles, Newton Abbot, 283pp.

Duff, K. L., McKirdy, A. P., & Harley, M. J. (eds.), 1985. *New sites for old: a students guide to the geology of the east Mendips*. Nature Conservancy Council, 189pp.

Edmonds, E. A., McKeown, M. C., & Williams, M., 1969. *British Regional Geology - South-West England* (3rd Ed.). HMSO, London, 130pp.

Findlay, D. C., Hawkins, A. B., & Lloyd, C. R., 1972. A gravel deposit on Bleadon Hill, Mendip, Somerset. *Proceedings of the University of Bristol Spelaeological Society*, **13**, 83-87.

Ford, D. C., & Stanton, W. I., 1968. The geomorphology of south-central Mendip Hills. *Proceedings of the Geologists' Association*, **79**, 401-427.

Gilbertson, D. D., 1979. The Burtle Sand Beds of Somerset: the significance of the freshwater interglacial molluscan faunas. *Proceedings of the Somerset Archaeology and Natural History Society*, **123**, 115-117.

Gilbertson, D. D., & Hawkins, A. B., 1974. Upper Pleistocene deposits and landforms at Holly Lane, Clevedon, Somerset (ST 419 727). *Proceedings of the University of Bristol Spelaeological Society*, **13**, 349-360.

Gilbertson, D. D., & Hawkins, A. B., 1977. The Quaternary deposits at Swallow Cliff, Middlehope, County of Avon. *Proceedings of the Geologists' Association,* **88**, 255-266.

Gilbertson, D. D., & Hawkins, A. B., 1978. The Col-Gully and glacial deposits at Court Hill, nr. Bristol, England. *Journal of Glaciology*, **20**, 173-188.

Gilbertson, D. D., & Hawkins, A. B., 1978. The Pleistocene succession at Kenn, Somerset. *Bulletin of the Geological Survey of Great Britain*, No. 66.

Gilbertson, D. D., & Mottershead, D. N., 1975. The Quaternary deposits at Doniford, west Somerset. *Field Studies*, **4**, 117-129.

Goldberg, P., & Macphail, R. I., 1990. Micromorphological evidence of Middle Pleistocene landscape and climatic changes from southern

England: Westbury-sub-Mendip, Somerset and Boxgrove, W Sussex. In: Douglas, L.A. (ed.) *Soil micromorphology*. Proc., 8th meeting of soil micromorphology, San Antonio, 1988. (Elsevier; Developments in Soil Science, 19), pp 441-447.

Green, G. W., & Donovan, D. T., 1969. The Great Oolite of the Bath area. *Bulletin of the Geological Survey of Great Britain*, No. 30, 1-63.

Green, G. W., & Welch, F. B. A., 1965. Geology of the country around Wells and Cheddar. *Memoirs of the Geological Survey of Great Britain*, No. 280, 225pp.

Greenly, E., 1919. The Pleistocene formations of Claverham and Yatton. *Proceedings of the Bristol Naturalists Society*, **5**, 145-147.

Greenly, E., 1922. An aeolian deposit at Clevedon. *Geological Magazine*, **59**, 365-376.

Hardy, P., 1999. *The Geology of Somerset*. Ex Libris Press, Bradford-on-Avon, 223pp.

Harrison, S., Anderson, E., & Passmore, D. G., 1998. A small glacial cirque basin on Exmoor, Somerset. *Proceedings of the Geologists' Association*, **109**, 149-158.

Haslett, S. K., 2001. Quaternary geology of the Severn Basin. *Proceedings of the Cotteswold Naturalists' Field Club*, **42** (1), 38-40.

Haslett, S. K. (editor), 2002. *Quaternary Environmental Micropalaeontology*. Arnold, London, 333pp.

Haslett, S. K., 2003. Foraminifera analysis of the Middle Jurassic reptile bearing deposits of Hornsleasow Quarry, Gloucestershire: a biostratigraphic and palaeoenvironmental analysis. *Proceedings of the Cotteswold Naturalists' Field Club*, **42** (3), 187-190.

Haslett, S. K., 2006. Topographic variation of an estuarine salt marsh: Northwick Warth (Severn Estuary, UK). *Bath Spa University Occasional Papers in Geography*, **3**, 1-17.

Haslett, S. K., 2008a. *Coastal Systems (2nd Edition)*. Routledge, London and New York, 240pp.

Haslett, S. K., 2008b. 400 Years On! Report of a public conference commemorating the 400th anniversary of the 1607 flood in the Bristol Channel and Severn Estuary, UK. *Archaeology in the Severn Estuary*, **18** (for 2007), 115-118.

Haslett, S. K., & Bryant, E. A., 2005. The AD 1607 coastal flood in the Bristol Channel and Severn Estuary: historical records from Devon and Cornwall. *Archaeology in the Severn Estuary*, **15** (for 2004), 81-89.

Haslett, S. K., & Bryant, E. A., 2007. Reconnaissance of historic (post-AD 1000) high-energy deposits along the Atlantic coasts of southwest Britain, Ireland and Brittany, France. *Marine Geology*, **242**, 207-220.

Haslett, S. K. & Bryant, E. A., 2008. Historic tsunami in Britain since AD 1000: a review. *Natural Hazards and Earth System Sciences*, **8**, 587-601.

Haslett, S. K., & Davies, P., 2002. Holocene lithostratigraphy and coastal change in the Somerset Levels: evidence from Nyland Hill, Axe Valley, Somerset. *Bath Spa University College Occasional Papers in Geography*, **2**, 37-43.

Haslett, S., Davies, P., Margetts, A., & Ford, T., 1996. Holocene coastal palaeoenvironments of the Axe Estuary, Somerset. In: M. G. Healy (ed.) *Late Quaternary coastal change in west Cornwall, UK: field guide*. Environmental Research Centre, Department of Geography, University of Durham, Research Publication 3, p. 109.

Haslett, S. K., Davies, P., Curr, R. H. F., Davies, C. F. C., Kennington, K., King, C. P., & Margetts, A. J., 1998. Evaluating late-Holocene relative sea-level change in the Somerset Levels, southwest Britain. *The Holocene*, **8**, 197-207.

Haslett, S. K., Davies, P., & Strawbridge, F., 1998. Reconstructing Holocene sea-level change in the Severn Estuary and Somerset Levels: the foraminifera connection. *Archaeology in the Severn Estuary,* **8**, 29-40.

Haslett, S. K., Davies, P., Davies, C. F. C., Margetts, A. J., Scotney, K. H., Thorpe, D. J., & Williams, H. O., 2001. The changing estuarine environment in relation to Holocene sea-level and the archaeological implications. *Archaeology in the Severn Estuary*, **11** (for 2000), 35-53.

Haslett, S. K., Strawbridge, F., Martin, N. A., & Davies, C. F. C., 2001. Vertical saltmarsh accretion and its relationship to sea-level in the Severn Estuary, UK: an investigation using Foraminifera as tidal indicators. *Estuarine, Coastal and Shelf Science*, **52**, 143-153.

Haslett, S. K., Davies, P., Eales, C. P., Vowles, E. M., & Williams, H. O., 2006. Variability in the Holocene lithostratigraphy of the Somerset Levels, UK. In: C. O. Hunt & S. K. Haslett (eds) *The Quaternary of Somerset: field guide*. Quaternary Research Association, Cambridge, 44-52 (refs 205-236).

Haslett, S. K., Howard, K. L., Margetts, A. J., & Davies, P., 2001. Holocene stratigraphy and evolution of the northern coastal plain of the

Somerset Levels, UK. *Proceedings of the Cotteswold Naturalists' Field Club*, **42** (1), 78-88.

Hawkins, A. B., 1962. The buried channel of the Bristol Avon. *Geological Magazine*, **99**, 369-374.

Hawkins, A. B., 1966. The geology of the Keynsham Bypass. *Proceedings of the Bristol Naturalists Society*, **31**, 195-202.

Hawkins, A. B., 1967. The geology of the Portbury area. *Proceedings of the Bristol Naturalists Society*, **31**, 421-428.

Hawkins, A. B., 1971. Sea level changes around south west England. *Colston Papers*, **23**, 67-88.

Hawkins, A. B., 1972. Some gorges of the Bristol District. *Proceedings of the Bristol Naturalists Society*, **32**, 167-185.

Hawkins, A. B., 1973. The geology of the slopes of the Bristol region. *Quarterly Journal of Engineering Geology*, **6**, 185-205.

Hawkins, A. B., & Kellaway, G. A., 1971. Field meeting at Bristol and Bath with special reference to new evidence of glaciation. *Proceedings of the Geologists' Association*, **82**, 267-291.

Hepworth, J. V., & Stride, A. H., 1950. A sequence from the Old Red Sandstone to Lower Carboniferous, near Burrington, Somerset. *Proceedings of the Bristol Naturalists Society*, **28**, 135-138.

Hinton, M. A. C., 1907. Note of the occurrence of the Alpine Vole (*Microtus nivalis*) in the Clevedon bone deposit. *Proceedings of the Bristol Naturalists Society*, **1**, 190-191.

Hollingworth, N. T. J., Ward, D. J., Simms, M. J., & Clothier, P., 1990. A temporary exposure of Lower Lias (Late Sinemurian) at Dimmer Camp, Castle Cary, Somerset, south-west England. *Mesozoic Research*, **2**, 163-180.

Holloway, S., & Chadwick, R. A., 1984. The I. G. S. Bruton Borehole (Somerset, England) and its regional structural significance. *Proceedings of the Geologists' Association*, **95**, 165-174.

Hunt, C. O. & Haslett, S. K. (eds), 2006a. *Quaternary of Somerset: field guide*. Quaternary Research Association, London, 236pp.

Hunt, C. O. & Haslett, S. K., 2006b. Quaternary Research Association Annual Field Meeting: Somerset and International Quaternary Union (INQUA) Field Meeting "Short term sea-level change and coastal vulnerability" project. In: C. O. Hunt & S. K. Haslett (eds) *The Quaternary of Somerset: field guide*. Quaternary Research Association, Cambridge, 1-12 (refs 205-236).

Ivimey-Cook, H. C., 1993. Lower Jurassic fossils from ST 63 NW, ST 64 SW and ST 54 SE, Glastonbury. *British Geological Survey Technical Report, WH/93/104R.*

Jarzembowski, E. A., 1989. Writhlington Geological Nature Reserve. *Proceedings of the Geologists' Association,* **100**, 219-234.

Jennings, S., Orford, J. D., Canti, M., Devoy, R. J. N., & Straker, V., 1998. The role of relative sea-level rise and changing sediment supply on Holocene gravel barrier development: the examples of Porlock, Somerset, UK. *The Holocene,* **8**, 165-181.

Jefferies, R. L., Willis, A. J., & Yemm, E. W., 1968. The Late and Post-Glacial history of the Gordano Valley, north Somerset. *New Phytologist,* **67**, 335-348.

Jones, J., Tinsley, H., McDonnell, R., Cameron, N., Haslett, S., & Smith, D., 2005. Mid Holocene coastal environments from Minehead Beach, Somerset, UK. *Archaeology in the Severn Estuary,* **15** (for 2004), 49-69.

Jones, M. S., Horbury, A., & Thompson, G. E., 1998. Characterization of freshly quarried and decayed Doulting limestone. Quarterly Journal of Engineering Geology, **31**, 325-331.

Kalaugher, P., & Grainger, P., 1996. Photographic monitoring of cliff recession at Watchet, Somerset. *Proceedings of the Ussher Society,* **9**, 17-21.

Kellaway, G. A., & Welch, F. B. A., 1948. *British Regional Geology - Bristol and Gloucester District. HMSO,* London, 91pp.

Kellaway, G. A., & Welch, F. B. A., 1993. *Geology of the Bristol District.* Memoir of the British Geological Survey, HMSO.

Kellaway, G. A., & Wilson, V., 1941. An outline of the geology of Yeovil, Sherborne and Sparkford Vale. *Proceedings of the Geologists' Association,* **52**, 131-174.

Kelly, P. G., Sanderson, D. J., & Peacock, D. C. P., 1998. Linkage and evolution of conjugate strike-slip fault zones in limestones of Somerset and Northumbria. *Journal of Structural Geology,* **20**, 1477-1493.

Kelly, P. G., Peacock, D. C. P., Sanderson, D. J., & McGurk, A. C., 1999. Selective reverse-reactivation of normal faults, and deformation around reverse-reactivated faults in the Mesozoic of the Somerset coast. *Journal of Structural Geology,* 21, 493-509.

King, A., 1997. *Fossil ammonites from the Somerset coast.* Somerset County Museum.

King, C., & Haslett, S., 1999. Modern saltmarsh foraminifera distribution in Stert Flats, Bridgwater Bay, UK: preliminary results. *Archaeology in the Severn Estuary*, **9** (for 1998), 92-94.

Macfadyen, W. A., 1970. *Geological highlights of the West Country*. Butterworths, London, 296pp.

Matthews, S. C., Butler, M., & Sadler, P. M., 1973. Lower Carboniferous successions in north Somerset; report by directors of field meeting. *Proceedings of the Geologists Association*, **84**, 175-179.

Mayes, J., 1998. Orographic influences on local weather: a Bristol Channel case study. *Geography*, **83**, 322-330.

Melville, R. V., & Freshney, E. C., 1982. *British Regional Geology - The Hampshire Basin and adjoining areas*. HMSO, London, 146pp.

Moore, L. R., & Trueman, A. E., 1937. The Coal Measures of Bristol and Somerset. *Quarterly Journal of the Geological Society, London*, **93**, 195-240.

Morgan, C. L., & Reynolds, S. H., 1904. The igneous rocks associated with the Carboniferous Limestone of the Bristol district. *Quarterly Journal of the Geological Society, London,* **60**, 137-157.

Page, H., 1982. Some notes on the geomorphological and vegetational history of the saltings at Brean. *Proceedings of the Somerset Archaeology and Natural History Society*, **126**, 119-125.

Page, K., 1996. Observations on the succession of ammonite faunas in the Bathonian (Middle Jurassic) of south-west England and their correlation with sub-Mediterranean 'Standard Zonation'. *Proceedings of the Ussher Society*, **9**, 45-53.

Palmer, C. P., 1972. The Lower Lias (Lower Jurassic) between Watchet and Lilstock in North Somerset (United Kingdom). *Newsletters on Stratigraphy*, **2**, 1-30.

Palmer, L. S., 1934. Some Pleistocene breccias near the Severn Estuary. *Proceedings of the Geologists Association*, **45**, 145-161.

Palmer, L. S., & Hinton, M. A., 1929. Some gravel deposits at Walton, near Clevedon. *Proceedings of the University of Bristol Spelaeological Society*, **3**, 154-161.

Parrott, R., 1976. The natural history of some eroded slopes on Compton Dundon escarpment. *Proceedings of the Somerset Archaeology and Natural History Society*, **120**, 21-28.

Parsons, C. F., 1975. Ammonites from the Doulting conglomerate bed (Upper Bajocian, Jurassic) of Somerset. *Palaeontology*, **18**, 191-205.

Peacock, D. C. P., & Sanderson, D. J., 1999. Deformation history and basin-controlling faults in the Mesozoic sedimentary rocks of the Somerset coast. *Proceedings of the Geologists' Association*, **110**, 41-52.

Penn, I. E., & Wyatt, R. J., 1979. The stratigraphy and correlation of the Bathonian strata in the Bath-Frome area. In: *The Bathonian Strata of the Bath-Frome Area*. Report of the Institute of Geological Sciences, No. 78/22, 23-88.

Pick, M. C., 1964a. The stratigraphy and sedimentary features of the Old Red Sandstone, Portishead coastal section, N. E. Somerset. *Proceedings of the Geologists' Association*, **75**, 199-221.

Pick, M. C., 1964b. The Triassic Dolomitic Conglomerate and structure of the Old Red Sandstone, Portishead coastal section, N. E. Somerset. *Proceedings of the Bristol Naturalists Society*, **30**, 445-450.

Proctor, C. J., 1994. Carboniferous fossil plant assemblages and palaeoecology at the Writhlington Geological Nature Reserve. *Proceedings of the Geologists' Association*, **105**, 277-286.

Proctor, C. J., 1998. Arthropleurids from the Westphalian D of Writhlington Geological Nature Reserve, Somerset. *Proceedings of the Geologists' Association*, **109**, 93-98.

Proctor, C., 1999. An Upper Carboniferous eurypterid from the Writhlington Geological Nature Reserve. *Proceedings of the Geologists' Association*, **110**, 263-265.

Prudden, H. C., 1993. *Geological trails in Yeovil*. South Somerset District Council, 14pp.

Prudden, H. C., 1995. *Ham Hill: its rocks and quarries*. South Somerset District Council, 16pp.

Prudden, H. C., 1999. Soil erosion in Somerset. *Nature in Somerset*, 29-31.

Pye, K., & Mottershead, D. N., 1995. Honeycomb weathering of Carboniferous sandstone in a sea wall at Weston-super-Mare, UK. *Quarterly Journal of Engineering Geology*, **28**(4), pp 333-347.

Ramsbottom, W. H. C., 1970. Carboniferous faunas and palaeogeography of the southwest England region. *Proceedings of the Ussher Society*, **2**, 144-157.

Reynolds, S. H., 1907a. A bone cave at Walton, near Clevedon. *Proceedings of the Bristol Naturalists Society*, **1**, 183-187.

Reynolds, S. H., 1907b. A Silurian inlier in the eastern Mendips. *Quarterly Journal of the Geological Society, London*, **63**, 217-240.

Reynolds, S. H., 1908. The igneous rocks of the Bristol district. *Proceedings of the Geologists' Association*, **20**, 59-65.

Reynolds, S. H., 1912. Further work on the Silurian rocks of the eastern Mendips. *Proceedings of the Bristol Naturalists Society*, **3**, 76-82.

Reynolds, S. H., 1916a. Further work on the igneous rocks associated with the Carboniferous Limestone of the Bristol district. *Quarterly Journal of the Geological Society, London*, **72**, 23-42.

Reynolds, S. H., 1916b. Carboniferous Limestone Series of the area between Clifton and Clevedon. *Proceedings of the Bristol Naturalists Society*, **4**, 186-197.

Reynolds, S. H., 1921. *A geological excursion handbook for the Bristol district* (2nd Ed.). Arrowsmith, Bristol, 224pp.

Reynolds, S. H., 1929. The geology of the Bristol district. *Proceedings of the Geologists' Association*, **40**, 77-103.

Reynolds, S. H., & Greenly, E., 1923. The Old Red Sandstone and Carboniferous Limestone of the Portishead-Clevedon area. *Proceedings of the Bristol Naturalists Society*, **6**, 92-97.

Reynolds, S. H., & Greenly, E., 1924. The geological structure of the Clevedon-Portishead area. *Quarterly Journal of the Geological Society, London*, **80**, 447-467.

Reynolds, S. H., & Vaughan, A., 1911. Faunal and lithological sequences in the Carboniferous Limestone Series (Avonian) of Burrington Combe (Somerset). *Quarterly Journal of the Geological Society, London*, **67**, 342-392.

Richardson, L., 1901. Mesozoic geography of the Mendip Archipelago. *Proceedings of the Cotteswold Naturalists Field Club*, **14**, 59-69.

Richardson, L., 1906. On a section of Middle and Upper Lias rocks near Evercreech, Somerset. *Geological Magazine*, **3**, 368-369.

Richardson, L., 1907. The Inferior Oolite and contiguous deposits of the Bath Doulting district. *Quarterly Journal of the Geological Society, London*, **63**, 383-436.

Richardson, L., 1909. On the sections of Inferior Oolite and the Midford-Camerton section of the Limpley Stoke railway, Somerset. *Proceedings of the Geologists' Association*, **21**, 97-100.

Richardson, L., 1909. On some Middle and Upper Lias sections near Batcombe, Somerset. *Geological Magazine*, **6**, 540-542.

Richardson, L., 1910. The Inferior Oolite and contiguous deposits of the south Cotteswolds. *Proceedings of the Cotteswold Naturalists Field Club*, **17**, 63-136.

Richardson, L., 1911. The Rhaetic and contiguous deposits of West, Mid and part of East Somerset. *Quarterly Journal of the Geological Society, London*, **67**, 1-74.

Richardson, L., 1916. The Inferior Oolite and contiguous deposits of the Doulting-Milbourne Port District (Somerset*). Quarterly Journal of the Geological Society*, **71**, 473-519.

Rippon, S., with contributions from G. Aalbersberg, J. R. L. Allen, S. Allen, N. Cameron, C. Gleed-Owen, P. Davies, S. Hamilton-Dyer, S. Haslett, J. Heathcote, J. Jones, A. Margetts, D. Richards, N. Shiel, D. Smith, J. Smith, J. Timby, H. Tinsley, & H. Williams, 2000. The Romano-British exploitation of coastal wetlands: survey and excavation on the North Somerset Levels, 1993-7. *Britannia*, **31**, 69-200.

Sanders, W., 1841. Account of a raised sea-beach at Woodspring Hill, near Bristol. *Report of the British Association of the Advancement of Science*, **1840** (transactions section), 102-103.

Savage, R. J. G., (ed.) 1977. *Geological excursions in the Bristol District*. University of Bristol, 196pp.

Seavill, E. W., 1941. Geology of the Bristol District. *Proceedings of the Bristol Naturalists Society*, **9**, 264-274.

Simms, M. J., 1995. The geological history of the Mendip Hills and their margins. *Proceedings of the Bristol Naturalists' Society*, **55**, 113-134.

Stead, J. T. G., & Williams, B. P., 1973. The Pennant Sandstone of Portishead. *Proceedings of the Bristol Naturalists Society*, **32**, 307-314.

Strawbridge, F., Haslett, S. K., Koh, A., Edwards, E., & Davies, C. F. C., 2000. The potential of Aerial Digital Photography for saltmarsh monitoring. *Bath Spa University College Occasional Papers in Geography*, **1**, 15pp + 4 plates.

Swift, A., 1995. A review of the nature and outcrop of the 'White Lias' facies of the Langport Member (Penarth Group: Upper Triassic) in Britain. *Proceedings of the Geologists' Association*, **106**(4), 247-258.

Sylvester-Bradley, P. C., & Hodson, F., 1957. The Fuller's Earth of Whatley, Somerset. *Geological Magazine*, **94**, 312-325.

Torrens, H. S., 1969. *International Field Symposium on the British Jurassic. Part B, Excursion No. 2, Guide to North Somerset*, pp. 2-26.

Trimmer, J., 1853. On the southern termination of the erratic Tertiaries, and on the remains of a bed of gravel on the summit of Clevedon Down, Somersetshire. *Quarterly Journal of the Geological Society, London*, **9**, 282-286.

Tutcher, J. W., & Trueman, A. E., 1925. The Liassic rocks of the Radstock District (Somerset). *Quarterly Journal of the Geological Society, London*, **81**, 595-666.

van de Kamp, P. C., 1969. The Silurian volcanic rocks of the Mendip Hills, Somerset; and the Tortworth area, Gloucestershire, England. *Geological Magazine*, **106**, 542-553.

Vaughan, A., 1905. The palaeontological sequence in the Carboniferous Limestone of the Bristol area. *Quarterly Journal of the Geological Society, London*, **61**, 181-307.

Wallis, F. S., 1927a. The Old Red Sandstone of the Bristol district. *Quarterly Journal of the Geological Society, London*, **83**, 760-789.

Wallis, F. S., 1927b. Notes on sections of Old Red Sandstone in the Bristol district. *Proceedings of the Bristol Naturalists Society*, **6**, 400-405.

Warrington, G., & Ivimey-Cook, H. C., 1995. The Late Triassic and Early Jurassic of coastal sections in west Somerset and South and Mid-Glamorgan. In: P. D. Taylor (ed.) *Field Geology of the British Jurassic*. Geological Society, London, pp. 9-30.

Webby, B. D., 1965a. The stratigraphy and structure of the Devonian rocks in the Brendon Hills, west Somerset. *Proceedings of the Geologists' Association*, **76**, 39-60.

Webby, B. D., 1965b. The Middle Devonian marine transgression in north Devon and west Somerset. *Geological Magazine*, **102**, 478-488.

Webby, B. D., 1965c. The stratigraphy and structure of the Devonian rocks in the Quantock Hills, west Somerset. *Proceedings of the Geologists' Association*, **76**, 321-344.

Webby, B. D., & Thomas, J. M., 1965. Whitsun field meeting: Devonian of west Somerset and Carboniferous of north-east Devon. *Proceedings of the Geologists' Association*, **76**, 179-194.

Wedlake, A. L., & Wedlake, D. J., 1963. Some palaeoliths from the Donniford gravels on the coast of west Somerset. *Proceedings of the Somerset Archaeology and Natural History Society*, **107**, 93-100.

Welch, F. B. A., 1929. The geological structure of the central Mendips. *Quarterly Journal of the Geological Society, London*, **85**, 45-76.

Welch, F. B. A., 1933. The geological structure of the eastern Mendips. *Quarterly Journal of the Geological Society, London*, **89**, 14-52.

Welch, F. B. A., 1956. Note on gravels at Kenn, Somerset. *Proceedings of the University of Bristol Spelaeological Society*, **7**, p. 137.

Whittaker, A., 1972a. The Watchet Fault - a post-Liassic transverse reverse fault. *Bulletin of the Geological Survey of Great Britain*, No. 41, 75-80.

Whittaker, A., 1972b. Account of an excursion to the west Somerset coast. *Report and Transactions of the Devonshire Association for the Advancement of Science*, **104**, 200-203.

Whittaker, A., 1973. The central Somerset Basin. *Proceedings of the Ussher Society*, **2**, 585-592.

Willemse, E. J. M., Peacock, D. C. P., & Aydin, A., 1997. Nucleation and growth of strike-slip faults in limestones from Somerset, UK. *Journal of Structural Geology*, **19**, 1461-1477.

Wilson, V., Welch, F. B. A., Robbie, J. A., & Green, G., 1958. Geology of the Country around Bridport and Yeovil. *Memoir of the Geological Survey of Great Britain, Sheets 312 and 327 (England and Wales)*.

Woodward, H. B., 1876. Geology of east Somerset and the Bristol coalfields. *Memoir of the Geological Survey of England and Wales*, 271pp.

Appendix - *Topographical and geological maps of Somerset*

The following lists give the names and sheet numbers of available geological and topographical 1:50,000 maps covering Somerset. Both are invaluable companions to the study and enjoyment of Somerset landscapes.

A.1 Topographical maps

The following list includes the Odnance Survey 1:50,000 Landranger Maps that cover Somerset:

Sheet number 171 - Cardiff and Newport
Sheet number 172 - Bristol, Bath and surrounding area
Sheet number 173 - Swindon and Devises, Marlborough and Trowbridge
Sheet number 180 - Barnstaple and Ilfracombe, Lynton and Bideford
Sheet number 181 - Minehead and Brendon Hills area
Sheet number 182 - Weston-super-Mare and Bridgwater area
Sheet number 183 - Yeovil and Frome
Sheet number 192 - Exeter, Sidmouth and surrounding area
Sheet number 193 - Taunton and Lyme Regis
Sheet number 194 - Dorchester, Weymouth and surrounding area

A.2 Geological maps

These are published by the British Geological Survey (formerly Institute of Geological Sciences) at a scale of 1:50,000, unless otherwise stated. Memoirs are available for many of these sheets. For details contact the British Geological Survey, Keyworth, Nottingham, NG12 5GG:

Sheet number 263 - Cardiff
Sheet number 264 - Bristol
Sheet number 265 - Bath (a 1:50,000 map enlarged from the 1:63,360 version)
Sheet number 278 and part of 294 - Minehead
Sheet number 279 and parts of 263 and 295 - Weston-super-Mare
Sheet number 280 - Wells
Sheet number 281 - Frome (scale 1:63,360)
Sheet number 294 - Dulverton
Sheet number 295 - Taunton
Sheet number 296 - Glastonbury
Sheet number 297 - Wincanton
Sheet number 310 - Tiverton
Sheet number 311 - Wellington
Sheet number 312 - Yeovil
Sheet number 313 - Shaftesbury

Made in the USA
Charleston, SC
08 March 2012